Lecture Notes in Mathematics 1675

Editors:
A. Dold, Heidelberg
F. Takens, Groningen

Springer
Berlin
Heidelberg
New York
Barcelona
Budapest
Hong Kong
London
Milan
Paris
Santa Clara
Singapore
Tokyo

Joseph E. Yukich

Probability Theory of Classical Euclidean Optimization Problems

 Springer

Author

Joseph E. Yukich
Department of Mathematics
Lehigh University
Bethlehem, PA 18015, USA
e-mail: jey0@lehigh.edu

Cataloging-in-Publication Data applied for

Die Deutsche Bibliothek - CIP-Einheitsaufnahme

Yukich, Joseph E.:
Probability theory of classical euclidean optimization problems / J. E.
Yukich. - Berlin ; Heidelberg ; New York ; Barcelona ; Budapest ;
Hong Kong ; London ; Milan ; Paris ; Santa Clara ; Singapore ;
Tokyo : Springer, 1998
  (Lecture notes in mathematics ; 1675)
  ISBN 3-540-63666-8

Mathematics Subject Classification (1991): 60C05, 60D05, 60F15, 60F10, 05C80

ISSN 0075-8434
ISBN 3-540-63666-8 Springer-Verlag Berlin Heidelberg New York

© Springer-Verlag Berlin Heidelberg 1998
Printed in Germany

Typesetting: Camera-ready T$_E$X output by the author
SPIN: 10553398   46/3143-543210 - Printed on acid-free paper

To Rahul, Nicholas, and Arati

PREFACE

This monograph aims to develop the probability theory of solutions to the classic problems in Euclidean combinatorial optimization, computational geometry, and operations research. These problems are naturally associated with graphs and our chief goal is to describe the almost sure (a.s.) behavior of the total edge length of these graphs. We do this by formulating a general approach which describes the total edge length behavior of graphs on random point sets in Euclidean space.

Many random Euclidean graphs, especially those motivated by some of the classic problems of discrete mathematics, have intrinsic similarities including self-similarity, subadditivity, and superadditivity. We use these similarities to prove laws of large numbers, rates of convergence, and large deviation principles for the solutions of the classic problems in combinatorial optimization. We prove limit theorems for the length of the shortest tour on a random sample, the minimal length of a tree spanned by a random sample, and the length of a minimal Euclidean matching on a random sample. The general tools and methods used to analyze these archetypical problems may also be used to study the stochastic behavior of geometric location problems, Steiner minimal spanning tree problems, and semi-matching problems.

The approach is not limited to problems in combinatorial optimization, but also provides the asymptotics for the total edge lengths of some of the fundamental graphs in computational geometry, including the k nearest neighbors graph. It is anticipated that the approach treats the lengths of the Voronoi and Delaunay tessellations of a random sample. The approach also yields asymptotics for graphs occurring in minimal surfaces, including the length of the minimal triangulation of a random sample as well as the area of the minimal tetrahedralization of a random sample, sometimes known as the probabilistic Plateau problem.

While the main goal is to develop a general structure describing the limit behavior of solutions to problems in combinatorial optimization, operations research, and computational geometry, there are also some secondary goals. First, we want to survey the remarkable progress in the field, much of which originates in the work of Steele, Rhee, and Talagrand. Along the way we describe some of the main open problems.

Second, we introduce a set of tools which have universal interest and which have applications beyond those mentioned here. The chief tool involves the simultaneous use of geometric subadditivity and superadditivity in nearly all problems. Additional useful tools include isoperimetry and martingale inequalities.

Most of the tools and methods come from probability and combinatorics. This monograph, which is essentially self-contained, may be read by both probabilists and combinatorialists. I have tried to make the monograph accessible to both sets of researchers and hope that it will benefit graph theorists and theoretical computer scientists. The close connection to problems of statistical mechanics may interest researchers in that field as well.

There is considerable overlap between the existing literature and this monograph. However, some of the foundational results, including Theorems 4.1 and 4.3, are new. The analysis of partitioning heuristics (Chapter 5), large deviations, and the applications of isoperimetry (Chapter 6) are also new. In Chapter 7 the generalized umbrella theorem for Euclidean functionals on $\mathbb{R}^d$ has not appeared before. Several of the applications in Chapter 8 are new, particularly the applications to the k nearest neighbors problem, the many traveling salesman problem, and the geometric location problem.

I owe a considerable debt of gratitude to Michel Talagrand, who at the outset encouraged me to write this monograph. His continued support, inspiration, and suggestions were invaluable. I am especially happy to thank Michael Steele for insightful suggestions and conversations covering a span of several years. His comments on a preliminary version of this book have improved the exposition in more ways than one. Numerous researchers and colleagues have provided valuable comments and it is a pleasure to acknowledge the assistance of Michael Aizenman, Amir Dembo, Vladimir Dobrić, Bennett Eisenberg, Wei-Min Huang, Garth Isaak, David Johnson, Kuntal McElroy, Kate McGivney, Charles Redmond, WanSoo Rhee, Peter Shor, and Ofer Zeitouni. In particular, Sungchul Lee and Daniel Rose read through most chapters in their entirety, checking for accuracy. They are due special thanks.

CONTENTS

1. INTRODUCTION

1.1. Chief Goals

This monograph concerns the study of the lengths of graphs on random vertex sets in Euclidean space $\mathbb{R}^d$, $d \geq 2$. We are interested primarily in the lengths of those graphs representing the solutions to problems in Euclidean combinatorial optimization.

This work was originally motivated by a desire to understand the asymptotics of the solutions of the classic problems of Euclidean combinatorial optimization, including the well-known traveling salesman problem. Later, as the tools and methods were refined and generalized, it was realized that the approach was general enough to describe the behavior of a variety of Euclidean graphs, including random tessellations, triangulations, and geometric location problems. We will limit our discussion to the lengths of some of the better known Euclidean graphs, particularly those occurring naturally in combinatorial optimization, computational geometry, and operations research. We expect that the methods and approach here can be used to treat other Euclidean graphs.

Within combinatorial optimization, there are several graphs which are of central importance. Many problems in combinatorial optimization involve the construction of the shortest possible network of some kind; we will be interested in the total edge length of these networks. Typical problems include the following ones, which, for the sake of brevity, are stated here in only an informal manner. Here V is a *random* vertex set in Euclidean space.

(i) Traveling Salesman Problem (TSP). Find the length of the shortest closed path traversing each vertex in V exactly once.

(ii) Minimum Spanning Tree (MST). Find the minimal total edge length of a spanning tree through V.

(iii) Minimal Euclidean Matching. Find the minimal total edge length of a Euclidean matching of points in V.

These three archetypical problems are central to combinatorial optimization and operations research. Precise statements of these problems appear at the end of the chapter.

Within computational geometry, we are motivated by the following problems.

(i) Find the total edge length of the Voronoi and Delaunay tessellations of V.

(ii) Find the length of the k nearest neighbors graph on V.

The graphs defined by these problems are not only useful in the theory of Euclidean optimization (Preparata and Shamos, 1985), but they are also of widespread interest in the natural and social sciences (Okabe et al., 1992).

Within the theory of minimal surfaces, optimization problems involve lengths of graphs and, more generally, areas of surfaces. In two dimensions one such problem involves minimizing the sum of the lengths of the edges in a triangulation of V; in three dimensions the analogous problem involves minimizing the sum of the areas of the faces of a tetrahedralization of V.

Our central goal is to develop a general unifying approach which describes the stochastic behavior of the total edge length of a variety of graphs, including the above mentioned ones. This is the main focus of the monograph. Considerable effort has gone into understanding the total edge length behavior of Euclidean graphs; however, up until now different problems have relied upon different approaches. We formulate a single approach which is simple and yet general enough to treat a variety of graphs arising in diverse areas. The approach contains known results as special cases and also opens up the way to proving new results.

We have two secondary goals. The first is to provide an overview of recent progress and to review related work. We will not present a complete history of the subject but will emphasize that work which is most relevant. We also indicate to the researcher some of the main unsettled problems in the field.

The second goal is to provide a set of tools which should be useful for future endeavors. Chief among these tools is the combined use of geometric subadditivity and superadditivity. Although subadditive methods have been used since the seminal work of Beardwood, Halton, and Hammersley (1959), superadditivity has not been explicitly used before. What is new here and what is not well recognized, is that most graphs having a subadditive structure admit canonical modifications, called boundary graphs, which possess an intrinsic superadditive structure. This insight can be put to good use and serves as the starting point towards deepening our understanding of total edge lengths of Euclidean graphs. Boundary graphs provide the key conceptual and technical tool of this monograph. They allow us to systematically and simultaneously exploit the inherent subadditivity and superadditivity of many Euclidean graphs. The combined use of subadditivity and superadditivity yields *two-sided additivity* and this forms the core of our approach.

Although the main new tool is the use of boundary graphs, there are other tools as well. One such tool is isoperimetry, developed by Talagrand (1995, 1996a, 1996b) and Rhee (1993b). Isoperimetry has already been recognized for its mathematical depth and breadth and it comes as no surprise that it finds applications in our subject. Indeed, Talagrand (1995) and Steele (1997) have showed that isoperimetric methods are germane to several problems of geometric probability.

1.2. A Brief History

We now consider the background of our subject, sketch its historical development, and review recent progress. Much of the probability theory of combinatorial optimization has been heavily influenced by the work of Steele.

Beardwood, Halton, and Hammersley (1959) proved the following celebrated result, which serves as a starting point for Steele's work as well as for this monograph:

Theorem 1.1. Let X_i, $i \geq 1$, be independent and identically distributed random variables with values in the unit cube $[0,1]^d$, $d \geq 2$. Let $T(X_1, ..., X_n)$ denote the length of the shortest tour through $X_1, ..., X_n$. Then with probability one

$$(1.1) \qquad \lim_{n \to \infty} T(X_1, ..., X_n)/n^{(d-1)/d} = \alpha(d) \int_{[0,1]^d} f(x)^{(d-1)/d},$$

where f is the density of the absolutely continuous part of the law of X_1 and $\alpha(d)$ is a positive constant which depends only on d.

Beardwood, Halton, and Hammersley (1959) recognized that the methods used to prove Theorem 1.1 had the potential to treat various problems. They indicated that Theorem 1.1 holds for the minimal spanning tree and Steiner minimal spanning tree problems, but did not explicitly develop the limit theory. They also conjectured that their approach could be useful in the probabilistic versions of Plateau's problem, Douglas's problem, and other problems in minimal surfaces. *A central goal of this monograph is to show through simple and general methods that an unexpectedly large number of problems in Euclidean combinatorial optimization and computational geometry have solutions satisfying the limit law (1.1).*

The elegant limit law (1.1) has several implications. Jensen's inequality shows that the right side of (1.1) is largest when f is the density of the uniform distribution. Hence non-uniformity in the distribution of X will tend to decrease the total tour length $T(X_1, ..., X_n)$. Also, since every probability distribution is the sum of a singular and a continuous component, the singular component of a probability distribution contributes nothing to the limit in (1.1). As pointed out by Beardwood, Halton, and Hammersley (1959), it represents the extreme case of non-uniformity.

Simple scaling arguments show that Theorem 1.1 holds if the unit cube is replaced by an arbitrary compact subset K of $\mathbb{R}^d$. In particular if X_i, $i \geq 1$, are independent and identically distributed (i.i.d.) random variables with the uniform distribution on a set K of Lebesgue measure one, then the limit in (1.1) exists and equals $\alpha(d)$ almost surely (a.s.). In other words, we find the rather surprising result that the limit is independent of the shape of the compact set K. In the sequel we will review the numerical estimates for the constant $\alpha(d)$.

The landmark work of Beardwood, Halton, and Hammersley (1959) stood virtually by itself for over fifteen years. In an unrelated effort, Miles (1970) showed that the mean total edge lengths of planar tessellations, including the Voronoi and Delaunay tessellations, satisfy the limit law (1.1). Moreover, he explicitly computed values for the limiting constants. However, his results are limited to the homogeneous planar Poisson point process and do not approach (1.1) in generality.

Several years later Karp recognized that the a.s. asymptotics (1.1) could be used to deepen our understanding of the approximate algorithmic solutions to the TSP. In his seminal work, Karp (1976,1977) introduced for all $\epsilon > 0$ a partitioning heuristic for the TSP which runs in polynomial time and with probability one yields a tour whose length is within a factor of $1 + \epsilon$ of the minimal tour length. In this way he was the first to show that the stochastic version of an NP-complete problem a.s. has a polynomial time algorithm yielding a nearly optimal solution.

Soon after Karp's deep results, Papadimitriou (1978a) recognized that the proof of Theorem 1.1 could be modified to show a similar result for the minimal matching problem, at least in dimension $d = 2$. More generally, he showed that if a functional L satisfies four key combinatorial conditions then it also satisfies the asymptotics (1.1). Two of these conditions, subadditivity and superadditivity, foreshadow the approach taken here.

Steele (1981a) also approached the subject from a general point of view. He abstracted the properties used in the proof of Theorem 1.1 and laid out a set of sufficient conditions guaranteeing that a large class of functionals, termed Euclidean functionals, satisfy the asymptotics (1.1). Steele's conditions involve geometric subadditivity, scaling, monotonicity, and translation invariance and he recognized that they form the key ingredients to the proof of Theorem 1.1. Using these abstract properties, Steele (1981a) proved a general result and deduced Theorem 1.1 as a corollary. He also showed via this general result that the length of the Steiner minimal spanning tree satisfies (1.1).

Steele's (1981a) classic paper opened the way for further research and soon it was shown that several other problems in combinatorial optimization enjoyed asymptotics similar to (1.1). Most of the effort however was confined to the study of problems on uniform samples, which, while certainly the most interesting case, does not tell the whole story. In the uniform setting Steele (1982) showed that a version of the minimal triangulation problem satisfies (1.1). Later, Steele (1988) explicitly showed that the length of the minimal spanning tree satisfies (1.1); Steele's work covers the case of power-weighted edges as well.

Steele's work was followed by several similar results. Avis, Davis, and Steele (1988) showed that the greedy matching heuristic satisfies the asymptotics (1.1). Talagrand (1991) showed that in the case of uniform random variables, (1.1) holds for the directed TSP, a modification of the usual TSP where edges are assigned directions according to an independent coin-tossing scheme. Goemans and Bertsimas (1991) employed Steele's general approach to show that (1.1) holds for the Held-Karp relaxation of the TSP. Later, Steele (1992) showed that the limit (1.1) holds for the semi-matching problem.

Several related results followed. Jaillet (1992), guided by the general abstract properties put forth by Steele, found suboptimal rates of convergence in (1.1). Rhee (1993) showed that the limit (1.1) could be strengthened to complete convergence for many optimization problems, including the minimal matching problem. Alexander (1994) discovered an approach which yields rates of convergence for the archetypical problems of combinatorial optimization. His rates are optimal in some cases.

These research efforts collectively suggest that (1.1) is a special case of a general "umbrella theorem" which includes the above results as special cases. As already indicated, our main goal is to show that this is indeed so. Our umbrella result opens the door for proving limit theorems for a variety of graphs. The methods used to prove this general "umbrella theorem" provide asymptotics for the lengths of graphs on arbitrary sequences of i.i.d. random variables. The methods also give rates of convergence.

1.3. Methods

Optimization problems and related problems involving lengths of graphs have been customarily viewed as functionals defined on point sets in Euclidean space. This approach is traditional and not unnatural. However, it is also useful to view these problems as functionals defined on *pairs* (F, R), where F is a point set in $\mathbb{R}^d$ and R is a d-dimensional rectangle in $\mathbb{R}^d$. Thus it will be useful to write, for example, $T(F, R)$ for the functional denoting the shortest tour through the set $F \cap R$.

This viewpoint helps identify the intrinsic similarities of optimization problems, especially geometric subadditivity, geometric superadditivity, ergodicity, translation invariance, and scaling. When considered as functionals on pairs (F, R), many optimization problems become superadditive functionals over the collection of d-dimensional rectangles. We use superadditivity to capture ergodicity. This is done by drawing upon a multiparameter version of Kingman's famous subadditive ergodic theorem.

One of our central ideas is that many problems in combinatorial optimization and computational geometry are not only subadditive, but admit simple and natural modifications having a superadditive structure. These modifications are found by examining the "boundary problem" or "boundary functional", an idea articulated in Redmond's (1993) thesis. It is easily seen and well-known that many optimization problems and problems in graph theory suffer from boundary effects. This peculiarity is an annoying irritant at best. At worst, it can lead to nearly insurmountable technical and conceptual difficulties. It is somewhat of an irony that it is precisely the peculiar boundary behavior which produces the coveted superadditive structure.

Roughly speaking, if the functional $L(F, R)$ denotes the length of a graph on a vertex set $F \subset R$, then the boundary functional $L_B(F, R)$ will denote the canonical functional associated with $L(F, R)$ which treats the boundary of R as a single point. Edges in the graph given by $L_B(F, R)$ which lie on the boundary of R are assigned zero length. In the case of the TSP functional, this means that travel along the boundary is "free". This simple idea leads to the valuable superadditive relation

$$L_B(F \cap (R_1 \cup R_2), R_1 \cup R_2) \geq L_B(F \cap R_1, R_1) + L_B(F \cap R_2, R_2)$$

which holds for all disjoint rectangles R_1 and R_2 whose union is a rectangle. Superadditive relations for L_B, coupled with subadditive relations for L, lead to two-sided additivity estimates for L. This monograph will explore the multiple benefits of two-sided additivity.

Boundary functionals L_B have a second crucial property which lies at the heart of our approach: the boundary functional L_B is "close" to the standard functional L. "Close" essentially means that the a.s. stochastic behavior of L is governed by that of L_B. Thus, to determine the stochastic behavior of L on point sets of large cardinality, it is usually easier to understand the stochastic behavior of the canonical boundary functional L_B and then use the closeness of L and L_B to capture the stochastic behavior of L.

Boundary problems and boundary functionals are essential tools, but they are not the only ones. Other tools involve isoperimetric methods, which are often more powerful than martingale methods. In particular, the Rhee and Talagrand isoperimetric inequalities have considerably furthered our understanding of the lengths of graphs. Their work, described in Chapter 6, essentially shows that the a.s. behavior of optimization functionals is governed by the behavior of the mean of the functional. Thus, to prove a.s. limit theorems for optimization functionals, it will be enough to prove limit results for the *mean* of the functionals. This observation dramatically simplifies the proofs.

There are several contributions provided by the general approach taken here:

(i) We provide a set of theorems which unifies probabilistic limit results for solutions to problems in optimization and computational geometry. The approach contains the classic asymptotic results and it also furnishes asymptotics for related problems in geometric probability, thereby generalizing and extending (1.1).

(ii) The level of generality of our main results includes the case of graphs with power-weighted edges. It also treats the case of vertex sets with unbounded support.

(iii) We use the intrinsic superadditivity of boundary functionals to find rates of convergence in (1.1) in a simple and natural way.

(iv) Limit results hold in the sense of complete convergence, which is stronger than a.s. convergence and which is necessary for some problems of model generation.

(v) We use boundary functionals to formulate large deviation results for the total edge lengths of graphs.

There are other benefits to the approach taken here. We recall that Karp (1976, 1977) introduced a partitioning heuristic for the TSP which runs in polynomial time and which a.s. yields a tour whose length is within a factor of $1 + \epsilon$ of the minimal tour length. We will generalize and extend Karp's result in the following way: given an optimization functional L, we will show that boundary functionals help analyze the performance of partitioning heuristics L_H which are analogous to Karp's partitioning heuristic and which approximate L to within a factor of $1 + \epsilon$.

There is a final benefit to our approach. Our motivation for considering boundary functionals came from a desire to better understand the stochastic behavior of the classic Euclidean optimization problems. It is a pleasant surprise that boundary functionals are useful in the *deterministic* setting as well. More precisely, boundary functionals are a natural tool in the study of the *worst case* values of problems in combinatorial optimization and operations research. The worst case value of L, denoted by $L(n)$, is the largest value of $L(V)$, where V ranges over all subsets of $[0, 1]^d$ of size n. Worst case versions of L are purely deterministic objects and it is somewhat surprising that they may be successfully analyzed using the approach developed for the stochastic analysis. We will see in Chapter 11 that the asymptotic behavior of the worst case versions parallels that of the stochastic versions.

1.4. Definitions

We recall the formal definitions of some of the classic problems of combinatorial optimization. The first several chapters use these concrete problems to illustrate the general theory. We emphasize that the general theory applies to a large variety of problems in geometric probability, including those motivated by problems in operations research and computational geometry. Some of these applications are explored in Chapters 8, 9, 10, and 11.

Our three-part definition begins with the problem of finding the shortest tour through a vertex set. This is perhaps the most famous problem of combinatorial optimization. Throughout $V := \{x_1, ..., x_n\} \subset \mathbb{R}^d$, $d \geq 2$.

Definition 1.2.

(a) (Traveling salesman problem; TSP) A closed tour or closed Hamiltonian tour is a closed path traversing each vertex in V exactly once. For all $p > 0$, let $T^p(V)$ be the length of the shortest closed tour T on V with pth power weighted edges. Thus

$$T^p(V) := \min_T \sum_{e \in T} |e|^p,$$

where the minimum is over all tours T and where $|e|$ denotes the Euclidean edge length of the edge e. Thus,

$$T^p(V) := \min_\sigma \left\{ \|x_{\sigma(n)} - x_{\sigma(1)}\|^p + \sum_{i=1}^{n-1} \|x_{\sigma(i)} - x_{\sigma(i+1)}\|^p \right\},$$

where the minimum is taken over all permutations σ of the integers $1, 2, ..., n$. $T^1(V)$ is the length of the classic traveling salesman tour on V.

(b) (Minimum spanning tree; MST) For all $p > 0$, let $M^p(V)$ be the length of the shortest spanning tree on V with pth power weighted edges, namely

$$M^p(V) := \min_T \sum_{e \in T} |e|^p,$$

where the minimum is over all spanning trees T of the vertex set V.

(c) (Minimal matching) For all $p > 0$, the minimal matching on V with pth power weighted edges has length given by

$$S^p(V) := \min_\sigma \sum_{i=1}^{n/2} \|x_{\sigma(2i-1)} - x_{\sigma(2i)}\|^p,$$

where the minimum is over all permutations of the integers $1, 2, ..., n$. If n has odd parity, then the minimal matching on V is the minimum of the minimal matchings on the n distinct subsets of V of size $n - 1$.

The above definition tells us that T^p, M^p, and S^p represent the canonical length functionals associated with the graphs given by the TSP, MST, and minimal matching problems, respectively. Definition 1.2 considers problems of combinatorial optimization which are more general than the classic ones which restrict attention to the case $p = 1$. The case of power-weighted edges ($p > 1$) is not studied as much as the linear case $p = 1$.   The approach taken here treats the general case with no additional work and so we have deliberately chosen to work with these general definitions.

Terminology, Notes, and References

(i) When $p = 1$ we will abbreviate notation and write $T(V)$ for $T^1(V)$ with similar meanings for $M(V)$ and $S(V)$.

(ii) $\mathcal{F} := \mathcal{F}(d)$ denotes the finite subsets of $\mathbb{R}^d$ and $\mathcal{R} := \mathcal{R}(d)$ denotes the d-dimensional rectangles of $\mathbb{R}^d$. These are sets of the form $[x_1, y_1] \times [x_2, y_2] \times ... \times [x_n, y_n]$, where $x_i, y_i \in [-\infty, \infty]$.

(iii) Throughout we use the symbol C to denote a constant which may depend on d and p and whose value may vary at each occurrence.

(iv) The Euclidean TSP problem is NP-complete (see e.g. Garey and Johnson (1979)) and there is no known algorithm for solving the TSP in a time which grows polynomially with the size of the vertex set. On the other hand, to solve the MST problem $M^p(V)$, a greedy algorithm (join the nearest two points with an edge, then the next pair of nearest points, and so on, being sure not to form a circuit) will work in polynomial time. We refer to Kruskal (1956), Prim (1957), and Borůvka (1926).

Among the many methods for finding the length $S^1(V)$ of a Euclidean minimal matching, Edmond's (1965) algorithm is probably the best known. Its running time, while not optimal, is of the order of $(\text{card}(V))^3$. Tarjan (1983) has a fine exposition of algorithms for optimization problems.

(v) It is well-known that combinatorial optimization problems can be formulated as problems in statistical mechanics. See e.g. Mézard, Parisi, and Virasoro (1987). Minimal matchings occur naturally in statistical mechanics. Finding the ground state energy of Ising models is equivalent to finding a spin configuration which minimizes the Hamiltonian of the system. This amounts to finding a minimum matching between the so-called frustrated plaquettes (the weights may not be Euclidean and therefore may not satisfy the triangle inequality). See Bieche et al. (1980) and Barahona et al. (1982).

2. SUBADDITIVITY AND SUPERADDITIVITY

2.1. Geometric Subadditivity

Mathematics is filled with a range of tools which are at once simple and powerful. It is difficult to find a tool which surpasses subadditivity in terms of its combined simplicity and utility. Subadditive methods occupy a prominent position in this monograph and we begin by recalling the basic notions. Let x_n, $n \geq 1$, be a sequence of real numbers satisfying the "subadditive inequality"

(2.1) $$x_{m+n} \leq x_m + x_n \quad \text{for all } m, n \in \mathbb{N}.$$

We say that the sequence x_n, $n \geq 1$, is *subadditive*. Subadditive sequences are nearly additive in the sense that they satisfy the *subadditive limit theorem*

$$\lim_{n \to \infty} \frac{x_n}{n} = \alpha$$

where $-\infty \leq \alpha < \infty$ and moreover

$$\alpha = \inf\{\frac{x_m}{m} : m \geq 1\}.$$

This is a standard result and we refer to Hille (1948) for a proof. The additive inverse of a subadditive sequence is a superadditive sequence, that is one which satisfies $x_{m+n} \geq x_m + x_n$ for all m and n. Subadditive sequences enjoy prominent use in various settings, including percolation and ergodic theory. They express the subadditivity of functions defined on the parameter set of intervals in $\mathbb{R}^+$.

Many sequences arising in applications are not subadditive but are "approximately subadditive" in the sense that they satisfy a generalized subadditive inequality of the form

$$x_{m+n} \leq x_m + x_n + \Delta_{m+n},$$

where Δ_k, $k \geq 1$, is an appropriate sequence. If Δ_k, $k \geq 1$, does not grow too fast then the sequence x_n, $n \geq 1$, still satisfies the subadditive limit theorem.

One of our central insights is that many graphs have an intrinsic subadditive and superadditive structure with respect to the parameter set of d-dimensional rectangles in $\mathbb{R}^d$. Although the subadditive structure expresses the self-similarity properties of the graph and is thus based on geometry in d dimensions, the analytic inequalities describing the corresponding graph length can usually be converted into inequalities involving approximately subadditive sequences. The intrinsic subadditivity of the traveling salesman graph, which will soon be made precise, was recognized by Beardwood, Halton, and Hammersley (1959). This observation formed the starting point for the proof of Theorem 1.1.

There are several ways to tease out the subadditive structure of graphs. For example, many functionals L^p occurring naturally in optimization problems satisfy a simple subadditivity condition

(2.2) $$L^p(F \cup G) \leq L^p(F) + L^p(G) + C_1 t^p$$

for all finite sets F and G in $[0,t]^d$ where C_1 is a finite constant which may depend upon d and p. It is easy to check that the TSP, MST, and minimal matching functionals given by Definition 1.2 each satisfy simple subadditivity (2.2).

Simple subadditivity (2.2) is often stronger than what is needed. In most cases, it is sufficient to consider a modification of (2.2) which expresses an approximate subadditivity of L^p over the collection $\mathcal{R} := \mathcal{R}(d)$ of d-dimensional rectangles. If $R \in \mathcal{R}$ is partitioned into rectangles R_1 and R_2 then (2.2) gives for all finite sets F

$$L^p(F \cap R) \le L^p(F \cap R_1) + L^p(F \cap R_2) + C_1(\mathrm{diam}R)^p.$$

This condition expresses *geometric subadditivity* of L^p over rectangles. Unlike the subadditive relation (2.1), the subadditivity is only an approximate one and carries a correction term of $C_1(\mathrm{diam}R)^p$.

Writing $L^p(F, R)$ for $L^p(F \cap R)$, we henceforth view L^p as a function defined on pairs (F, R) where F is a finite set and $R \in \mathcal{R}$ is a d-dimensional rectangle. This notation is not accidental and in fact explicitly recognizes that L^p is a function of two arguments, a point of view which is central to the development of our subject. With this notation we obtain

$$(2.3) \qquad L^p(F, R) \le L^p(F, R_1) + L^p(F, R_2) + C_1(\mathrm{diam}R)^p.$$

Although we emphasize that (2.3) represents only approximate subadditivity, we will sometimes omit mention of "approximate" and simply refer to (2.3) as geometric subadditivity. The relation (2.3) may appear weak but we will see that when F represents a random set it can be turned into an inequality involving sequences which are approximately subadditive. This device will form the foundation of our approach.

By iterating (2.3) it is straightforward to check that if $\{Q_i\}_{i=1}^{2^{dj}}$ is a partition of $[0,1]^d$ into subcubes of edge length 2^{-j} then for $0 < p < d$ we have

$$(2.4) \qquad L^p(F, [0,1]^d) \le \sum_{i=1}^{2^{dj}} L^p(F \cap Q_i, Q_i) + C_1' 2^{(d-p)j},$$

where the constant C_1' depends now on d and p. Indeed, when $j = 1$ we obtain for the partition $\{Q_i\}_{i=1}^{2^d}$

$$L^p(F, [0,1]^d) \le \sum_{i=1}^{2^d} L^p(F \cap Q_i, Q_i) + C_1 2^d D^p,$$

where $D := \mathrm{diam}[0,1]^d = d^{1/2}$. When $j = 2$ we obtain

$$L^p(F, [0,1]^d) \le \sum_{i=1}^{2^{2d}} L^p(F \cap Q_i, Q_i) + \sum_{i=1}^{2^d} 2^d \cdot C_1 \cdot (D/2)^p +$$
$$+ C_1 2^d D^p$$
$$\le \sum_{i=1}^{2^{2d}} L^p(F \cap Q_i, Q_i) + C_1 2^d \cdot D^p (2^{d-p} + 1).$$

Iterating j times gives

$$L^p(F, [0,1]^d) \leq \sum_{i=1}^{2^{jd}} L^p(F \cap Q_i, Q_i) +$$

$$+ \; C_1 2^d \cdot D^p (2^{(j-1)(d-p)} + \dots + 2^{d-p} + 1)$$

which is exactly (2.4).

The subadditive relation (2.3) is simpler than the customary one appearing in problems of this sort. One usually requires that (2.4) hold whenever Q_i, $i \geq 1$, are congruent subcubes with edge length $m^{-1/d}$, so that the correction term on the right side of (2.4) is $C_1 m^{(d-p)/d}$. Checking such a condition is a little more involved than checking the relation (2.3). We have chosen the form (2.3) of subadditivity for three reasons: the first is that it will conveniently accommodate problems defined on domains larger than the unit cube, the second is that it will parallel an analogous definition of superadditivity, and the third is its evident simplicity.

Many of the classic optimization problems satisfy geometric subadditivity (2.3). For example, to see that the length of the minimal spanning tree M^p is subadditive (2.3) we argue as follows: given a finite set F and a rectangle $R := R_1 \cup R_2$, let T_i denote the minimal spanning tree which realizes $M^p(F \cap R_i, R_i)$, $1 \leq i \leq 2$. Tie together the local spanning trees T_1 and T_2 with an edge which has a length bounded by the sum of the diameters of the rectangles R_1 and R_2 (see Figure 2.1). Performing this operation generates a feasible tree on F at a total cost bounded by the right side of (2.3). Now (2.3) follows by minimality. Showing that the TSP and minimal matching problem satisfy (2.3) involves similar considerations.

Figure 2.1. The MST is subadditive: the length of the global MST is bounded by the lengths of the local trees T_1 and T_2 and a connecting edge E

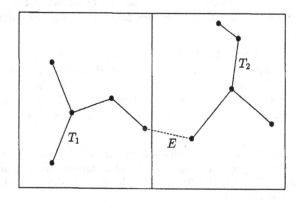

2.2. Geometric Superadditivity, Boundary Functionals

Subadditive relations (2.3) are significantly strengthened when coupled with superadditive relations. There are several reasons to search for superadditivity. The most compelling one is that if we can show superadditivity together with the approximate subadditivity (2.4), then the functional L^p becomes "nearly additive" in the sense that

$$L^p(F \cap R) \approx L^p(F \cap R_1) + L^p(F \cap R_2).$$

Relations of this sort are crucial in showing that a global graph length can be approximately expressed as a sum of the lengths of local components.

Unfortunately, most optimization functionals L^p lack an intrinsic superadditive property which leads to interesting limit results. This drawback motivated Steele (1981a) to define a condition which can occasionally serve as a substitute for superadditivity. His condition, termed "upper linearity", can be put to use in some situations.

The most convenient and elegant way to circumvent the lack of superadditivity involves an appropriate modification of the functionals L^p. The behavior of functionals on point sets in the unit cube is almost always influenced by the boundary of the unit cube. This presents annoying technical and conceptual difficulties and can sometimes be overcome by identifying the opposite faces of the cube. This amounts to replacing the usual metric on the cube by the "flat metric" and is usually an undesirable oversimplification.

We use this apparently annoying boundary effect to our advantage. We will use the boundary effects to modify the original problem into a "boundary problem". Boundary problems give rise to "boundary graphs", whose lengths we refer to as "boundary functionals". "Boundary functionals" will have precisely the sought after superadditivity. Roughly speaking, boundary functionals, denoted here by L_B, are defined on pairs $(F, R) \in \mathcal{F} \times \mathcal{R}$, and measure the length of those graphs on the vertex set F together with the boundary of R, which is treated as a single point. Thus edges on the boundary of R have zero length. In the case of the traveling salesman functional, this means that travel on the boundary ∂R is free. Implicit in their definition is the fact that boundary functionals L_B are smaller than the standard functional L.

The boundary functional $L_B(F, R)$ treats the boundary of R as a single point so that all edges joined to the boundary are joined to one another. Cognoscenti will recognize that boundary functionals are analogous to the "wired boundary condition" used in percolation and statistical mechanics. They are also analogous to the "wired spanning forest" used in the study of random trees (cf. Lyons and Peres, 1997).

As we will see shortly, boundary functionals L_B, which represent a slight modification of the underlying optimization functional L, are intrinsically superadditive. L is intrinsically subadditive. It follows that L is "nearly additive" whenever L_B is a "close" approximation to L.

This crucial feature makes boundary functionals a natural choice of study and, as we will see in the sequel, leads to a wealth of asymptotic estimates. In many instances, asymptotics for optimization functionals can be deduced from the study of asymptotics for boundary functionals. This idea is at the heart of our subject and will appear over and over again.

If $R \in \mathcal{R}$ is partitioned into rectangles R_1 and R_2 then a typical boundary functional L_B satisfies

$$(2.5) \qquad L_B(F, R) \geq L_B(F \cap R_1, R_1) + L_B(F \cap R_2, R_2)$$

and thus L_B is *superadditive with no error term*. The absence of an error term in (2.5) stands in sharp contrast to the approximate subadditive relation (2.3). In the sequel we will convert (2.5) into an inequality involving sequences which are superadditive and not merely "approximately superadditive". This distinction has telling consequences and will play a central role in our development of limit theorems for optimization functionals. This will be developed more fully in Chapter 4.

We illustrate the idea of boundary functionals by formally describing some concrete examples. Other examples will follow in the sequel.

The Boundary TSP Functional

For all d-dimensional rectangles $R \in \mathcal{R}$, finite sets $F \subset R$, and $p \geq 1$, let $T^p(F, R, (a, b))$ denote the length of the shortest path with pth power weighted edges through $F \cup \{a, b\}$ with endpoints a and b, where a and b belong to ∂R. Define the boundary functional T_B^p associated with T^p by

$$T_B^p(F, R) := \min \left(T^p(F, R), \ \inf \sum_i T^p(F_i, R, (a_i, b_i)) \right),$$

where the infimum ranges over all partitions $(F_i)_{i \geq 1}$ of F and all sequences of pairs of points $(a_i, b_i)_{i \geq 1}$ belonging to ∂R.

The boundary functional $T_B^p(F, R)$ may be interpreted as the length of an optimal cycle (with pth power weighted edges) through the set F which may repeatedly exit to the boundary of R at one point and re-enter at another, incurring no cost when moving along the boundary. See Figure 2.2. It is easy to check that T_B^p satisfies superadditivity (2.5): for each $1 \leq i \leq 2$, the restriction of the global tour $T_B^p(F, R)$ to the rectangle R_i defines a boundary tour of the set $F \cap R_i$, which by minimality is at least as large as $T_B^p(F, R_i)$. This is precisely (2.5). Superadditivity breaks down for the functional T_B^p when $0 < p < 1$.

Figure 2.2. The boundary TSP graph

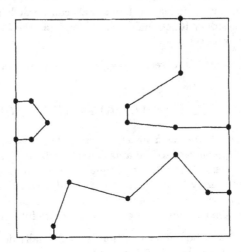

It is likewise easy to show that T_B^p satisfies simple subadditivity (2.2). The boundary TSP functional T_B^p is thus a modification of the standard TSP functional T^p. The boundary functional would not be so interesting were it not for its close approximation of the standard functional T^p. This closeness property is examined further in the next chapter.

The Boundary MST Functional

For all d-dimensional rectangles $R \in \mathcal{R}$, finite sets $F \subset R$, and $p \geq 1$, define the boundary MST functional by

$$M_B^p(F, R) := \min \left(M^p(F, R), \ \inf \sum_i M^p(F_i \cup a_i) \right),$$

where the infimum ranges over all partitions $(F_i)_{i \geq 1}$ of F and all sequences of points $(a_i)_{i \geq 1}$ belonging to ∂R. When $M_B^p(F, R) \neq M^p(F, R)$ the graph realizing the boundary functional $M_B^p(F, R)$ may be thought of as a collection of small trees connected via the boundary ∂R into a single large tree, where the connections on ∂R incur no cost. See Figure 2.3. It is a simple matter to see that the boundary MST functional is simply subadditive (2.2) and also superadditive (2.5). Later we will see that the boundary MST functional closely approximates the standard MST functional M^p.

Figure 2.3. The boundary MST graph

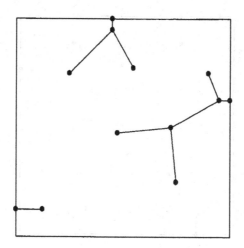

The Boundary Minimal Matching Functional

For all d-dimensional rectangles $R \in \mathcal{R}$, finite sets $F \subset R$, and $p \geq 1$, we let $S^p_B(F, R)$ denote the length of the least Euclidean matching (with pth power weighted edges) of points in F with matching to points on ∂R permitted. More precisely, each point in F is paired with either a boundary point on ∂R or another point in F; $S^p_B(F, R)$ minimizes the sum of the pth powers of the edge lengths over all such pairings. We allow the possibility that even when F has even parity, one point in F may be isolated and unmatched. This is needed to ensure superadditivity and to account for the case that minimal matching on an odd number of points leaves one point unmatched. See Figures 2.4 and 2.5.

As is the case with the boundary TSP and MST functionals, it is easy to verify that the boundary minimal matching functional is simply subadditive (2.2) and superadditive (2.5). See Figure 2.5. Later we will see that the boundary minimal matching functional closely approximates the standard minimal matching functional S^p.

Figure 2.4.   The boundary minimal matching graph

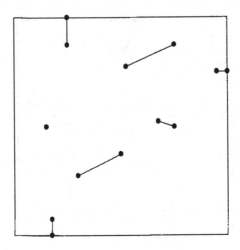

Figure 2.5.   Boundary minimal matching is superadditive: the length of the global matching (left) exceeds the length of the two local matchings (right)

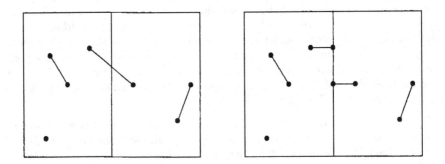

In the above examples, the boundary functional measures the minimal length of a graph satisfying a specified boundary condition. We have restricted attention to the range $p \geq 1$ since without further modifications the functionals would no longer enjoy superadditivity.

Notes and Notation

(i) If L is an optimization functional, then L_B will henceforth designate the canonical boundary functional. We will often abbreviate the notation and write $L^p(F)$ for $L^p(F, [0,1]^d)$ and similarly for $L_B^p(F)$. When $p = 1$ we will suppress mention of p and simply write $L(F, R)$ and $L_B(F, R)$ in place of $L^1(F, R)$ and $L_B^1(F, R)$, respectively. Notice that $L^p(F, \mathbb{R}^d) = L_B^p(F, \mathbb{R}^d)$.

Many of the subadditive functionals L^p considered in this monograph satisfy $L^p(F, R_1) = L^p(F, R_2)$ if $F \cap R_1 = F \cap R_2$. On the other hand, the superadditive functionals exhibit a higher dependence on the underlying rectangular domain and in general will *not* satisfy $L^p(F, R_1) = L^p(F, R_2)$ when $F \cap R_1 = F \cap R_2$.

(ii) We use L_B to designate the canonical boundary functional. Previous work of Redmond and Yukich (1994, 1996) referred to boundary functionals as "rooted duals" and denoted them by L_r.

3. SUBADDITIVE AND SUPERADDITIVE

EUCLIDEAN FUNCTIONALS

3.1. Definitions

In this chapter we lay the foundation for some of the basic limit results in geometric probability. As is often the case in mathematics, we search for a structure which is general enough to encompass a wide variety of problems, yet strong enough to yield results of interest. With this in mind, we formulate the idea of subadditive and superadditive Euclidean functionals and explore their crucial properties. The definitions and properties set forth in this chapter will be used over and over in the sequel.

We have previously seen that subadditive (respectively, superadditive) functionals $L^p(\cdot, R)$, $R \in \mathcal{R}$, enjoy a subadditive (respectively, superadditive) structure over the parameter set of d-dimensional rectangles. The following conditions endow the functional $L^p(F, R)$, $(F, R) \in \mathcal{F} \times \mathcal{R}$, with a *Euclidean structure* as well:

(3.1) $\qquad \forall\, y \in \mathbb{R}^d, \;\; R \in \mathcal{R}, \;\; F \subset R \quad L^p(F, R) = L^p(F + y, R + y)$

and

(3.2) $\qquad \forall\, \alpha > 0, \;\; R \in \mathcal{R}, \;\; F \subset R \quad L^p(\alpha F, \alpha R) = \alpha^p L^p(F, R).$

Conditions (3.1) and (3.2) express the *translation invariance* and *homogeneity of order p* of L^p, respectively.

It is easy to check that many functionals defined on point sets in Euclidean space satisfy translation invariance and homogeneity. This is especially true of functionals describing the length of a given graph. The TSP, MST, and minimal matching functionals are but a few examples. Functionals describing the length of a graph with pth power weighted edges are in general homogeneous of order p. These are not the only functionals which are homogeneous of order p. The probabilistic Plateau functional, essentially the minimal triangulation problem in three dimensions, does not describe edge lengths but instead describes surface area and in this way is homogeneous of degree 2. This is considered in Chapter 9.

If a functional $L^p(F, R)$, $(F, R) \in \mathcal{F} \times \mathcal{R}$, is superadditive over rectangles and has a Euclidean structure over $\mathcal{F} \times \mathcal{R}$, then we will say that L^p is a *superadditive Euclidean functional*. Formally we have the following definition, which is central to all that follows:

Definition 3.1. Let $L^p(\emptyset, R) = 0$ for all $R \in \mathcal{R}$ and suppose L^p satisfies (3.1) and (3.2). If L^p satisfies

(3.3) $$L^p(F, R) \geq L^p(F \cap R_1, R_1) + L^p(F \cap R_2, R_2),$$

whenever $R \in \mathcal{R}$ is partitioned into rectangles R_1 and R_2 then L^p *is a superadditive Euclidean functional of order p. Subadditive Euclidean functionals of order p satisfy* (3.1), (3.2), and geometric subadditivity

(3.4) $$L^p(F, R) \leq L^p(F \cap R_1, R_1) + L^p(F \cap R_2, R_2) + C_1(\operatorname{diam} R)^p.$$

Remarks.

(i) As noted in Chapter 2, if $\{Q_i\}_{i=1}^{2^{dj}}$ is a partition of $[0,1]^d$ into 2^{dj} subcubes of edge length 2^{-j}, then repeated applications of (3.4) yield for $0 < p < d$

(3.5) $$L^p(F, [0,1]^d) \leq \sum_{i=1}^{2^{dj}} L^p(F \cap Q_i, Q_i) + C_1 2^{(d-p)j},$$

for a new value of the constant C_1. We will make frequent use of this estimate.

(ii) We tacitly assume that $L^p(F, R)$ takes values in $(0, \infty)$ if $\operatorname{card} F > 1$. We will also assume that $L^p(F, R)$ is a measurable function from $(\mathbb{R}^d)^n$ to $[0, \infty)$, where $F \subset \mathbb{R}^d$ and $\operatorname{card} F = n$. This assumption is rather benign, since essentially all Euclidean functionals are continuous functions of the input F in the usual sense: small changes in F with respect to the Euclidean distance produce small changes in $L^p(F, R)$.

We will also tacitly assume the finiteness condition $\sup_{y \in [0,1]^d} L^p(\{y\}, [0,1]^d) < \infty$. This assumption is highly non-restrictive and is satisfied by essentially all Euclidean functionals.

(iii) The superadditive Euclidean functionals considered here are always the canonical boundary functionals associated with some standard Euclidean functional L. We therefore will henceforth designate superadditive Euclidean functionals as L_B.

It is not clear from the definition that there are many subadditive and superadditive Euclidean functionals. The following lemma, an immediate consequence of Chapter 2, tells us that the classic Euclidean optimization problems yield subadditive and superadditive Euclidean functionals.

Lemma 3.2. *For all $p > 0$, the functionals T^p, M^p, and S^p are subadditive Euclidean functionals of order p. For all $p \geq 1$, their respective boundary versions T_B^p, M_B^p, and S_B^p are superadditive Euclidean functionals of order p.*

In the remainder of this chapter we describe the salient features of subadditive and superadditive Euclidean functionals. These properties are critical and will be used throughout the monograph.

3.2. Growth Bounds

Geometric subadditivity (3.4) leads to several non-trivial consequences for the functional L^p. It is rather surprising that subadditivity leads to growth bounds for L^p. This observation was first noticed by Rhee (1993b) and has a wide range of applications. By using (3.5) with $j = 1$ and induction arguments, we may show the following growth estimate for L^p, which was first proved by Rhee (1993b) for the case $p = 1$.

Lemma 3.3. (growth bounds) Let L^p be a subadditive Euclidean functional of order p, $0 < p < d$. Then there exists a finite constant $C_2 := C_2(d, p)$ such that for all cubes R and all $F \subset R$ we have

$$(3.6) \qquad L^p(F, R) \leq C_2 (\operatorname{diam} R)^p (\operatorname{card} F)^{(d-p)/d}.$$

Proof. We follow Rhee (1993b). By homogeneity we may without loss of generality assume that R is the unit cube $[0, 1]^d$. If $\{Q_i\}_{i=1}^{2^d}$ is a partition of $[0, 1]^d$ into congruent subcubes of edge length $1/2$ then subadditivity (3.5) implies that

$$L^p(F, [0, 1]^d) \leq \sum_{i=1}^{2^d} L^p(F \cap Q_i, Q_i) + C_1',$$

where $C_1' := C_1 \cdot 2^{d-p}$. This simple subadditive estimate will be useful in the proof, which proceeds via induction on $\operatorname{card} F$.

To formulate the inductive proof, set $a := \sup_{y \in [0,1]^d} L^p(\{y\}, [0, 1]^d)$ and note that a is finite by assumption. Let $a_2 := \frac{C_1'}{2^{d-p}-1}$ and let $a_1 := a + \frac{d}{p} 2^{d-p} a_2$. We wish to show for all $F \subset [0, 1]^d$ that

$$L^p(F, [0, 1]^d) \leq a_1 (\operatorname{card} F)^{(d-p)/d}.$$

As for the induction hypothesis itself, we will assume that the stronger bound

$$L^p(F, [0, 1]^d) \leq a_1 (\operatorname{card} F)^{(d-p)/d} - a_2$$

holds whenever $\operatorname{card} F < n$. Note that $a \leq a_1 - a_2$ and so the induction hypothesis holds when $\operatorname{card} F = 1$.

Consider the partition $\{Q_i\}_{i=1}^{2^d}$ of $[0, 1]^d$. By assumption, F is not contained in any of the subcubes Q_i. Therefore for all $1 \leq i \leq 2^d$ we have $n_i := \operatorname{card}(F \cap Q_i) < n$ and thus by the induction hypothesis and homogeneity we have

$$L^p(F \cap Q_i, Q_i) \leq 2^{-p}(a_1 n_i^{(d-p)/d} - a_2).$$

By geometric subadditivity and the assumption that $L^p(\emptyset, R) = 0$ for all rectangles R we obtain

$$L^p(F, [0,1]^d) \leq 2^{-p} \sum_{i:n_i>0} (a_1 n_i^{(d-p)/d} - a_2) + C_1'.$$

Letting $m := \text{card}\{i : n_i > 0\}$, Hölder's inequality implies

$$L^p(F, [0,1]^d) \leq 2^{-p} a_1 m^{p/d} n^{(d-p)/d} - 2^{-p} a_2 m + C_1'$$
$$= 2^{-p} a_1 m^{p/d} n^{(d-p)/d} - 2^{-p} a_2 m + (2^{d-p} - 1) a_2,$$

where the last equality uses the definition of a_2.

It suffices to show that the right side of the above is at most $a_1 n^{(d-p)/d} - a_2$, or, equivalently, that

$$a_2(2^{d-p} - 2^{-p} m) \leq a_1 n^{(d-p)/d} \left(1 - 2^{-p} m^{p/d}\right).$$

Since $n \geq 1$ and $a_1 \geq \frac{d}{p} 2^{d-p} a_2$ it suffices to show

$$2^d - m \leq \frac{d}{p} 2^d \left(1 - 2^{-p} m^{p/d}\right).$$

If we choose x so that $m = x 2^d$ then this reduces to showing the inequality

$$1 - x \leq \frac{d}{p}(1 - x^{p/d})$$

for $0 < x \leq 1$ and $0 < p < d$. This last inequality follows from the observation that the graph of the function $\frac{d}{p}(1 - x^{p/d}) + x - 1$ has a minimum at $(1, 0)$. This verifies the induction step when $\text{card} F = n$ and completes the proof of Lemma 3.3. $\square$

For some Euclidean functionals, there are other ways to obtain growth bounds without appealing to Lemma 3.3. For example when $p = 1$, easy arguments based on the pigeonhole principle give a direct proof of (3.6) for the standard TSP T. When L^p is the power-weighted TSP functional T^p, the bound (3.6) is also a consequence of the space filling curve heuristic, as shown by Steele (1990). In fact, using this heuristic we may easily obtain for all $p > 0$ and $d \geq 1$ the growth bound

$$(3.7) \qquad L^p(F) \leq C \left((\text{card} F)^{(d-p)/d} \vee 1\right)$$

which is valid when L is either the TSP, MST, or minimal matching functional. (Here and elsewhere $x \vee y$ denotes the maximum of the real numbers x and y.) The interest of Rhee's bound (3.6) is that it holds for all subadditive Euclidean functionals of order p, $0 < p < d$.

The space filling curve heuristic is elegantly described by Steele (1997) and for the sake of completeness we briefly recall its applicability to the TSP. We follow the exposition of Steele (1997).

Let $\{x_1, ..., x_n\}$ be a point set in $[0,1]^d$, $d \geq 2$. To bound the length $T^p(x_1, ..., x_n)$ we consider a continuous function ϕ from $[0,1]$ *onto* $[0,1]^d$ that is Lipschitz of order $1/d$, that is for all $0 \leq s, t \leq 1$ we have

$$\|\phi(s) - \phi(t)\| \leq C|s - t|^{1/d}.$$

To find the length of a short feasible tour through $\{x_1, ..., x_n\}$ we use the space filling function ϕ and follow this recipe:

· compute a set of points $\{t_1, ..., t_n\} \subset [0,1]$ such that $\phi(t_i) = x_i$, $1 \leq i \leq n$,

· order the t_i so that $t_{(1)} \leq t_{(2)} \leq ... \leq t_{(n)}$, and

· define a permutation $\sigma : [1, n] \to [1, n]$ by requiring that $x_{\sigma(i)} = \phi(t_{(i)})$.

The feasible tour which visits $x_1, ..., x_n$ in the order of $x_{\sigma(1)}, x_{\sigma(2)}, ..., x_{\sigma(n)}$ satisfies the length estimate

$$\sum_{i=1}^{n-1} |x_{\sigma(i)} - x_{\sigma(i+1)}|^p = \sum_{i=1}^{n-1} |\sigma(t_{(i)}) - \sigma(t_{(i+1)})|^p$$

$$\leq C^p \sum_{i=1}^{n-1} |t_{(i)} - t_{(i+1)}|^{p/d}$$

$$\leq C^p(n^{1-p/d} \vee 1),$$

where for $0 < p < d$ we use Hölder's inequality and for $p \geq d$ we use the estimate $\sum |t_{(i)} - t_{(i+1)}|^{p/d} \leq \sum |t_{(i)} - t_{(i+1)}| \leq 1$. Thus

$$T^p(x_1, ..., x_n) \leq C \left(n^{(d-p)/d} \vee 1 \right)$$

as desired.

The space filling curve heuristic thus provides an easy way to obtain the growth bound (3.7) even for the delicate case $p \geq d$. Without the heuristic, the proof of the bound (3.7) for this last case is no easy task.

3.3. Smoothness

We will soon see that geometric superadditivity (3.3) and subadditivity (3.4) are powerful tools when used together. They gain further strength in the presence of smoothness conditions on the functional L^p, which describe the variation of L^p as points are added and deleted. Functionals which are monotone in the sense that $L(F) \leq L(F \cup x)$ for all point sets F and singletons x have variations which are easy to describe. This is the case with the TSP functional, for example. Most functionals are not so easy to describe, however, and this is where smoothness conditions become important. Smoothness, along with subadditivity and superadditivity, occupies a central role in framing the limit theorems in this monograph.

Many Euclidean functionals enjoy the following smoothness condition, forms of which were used by Steele (1988) and Avis, Davis, and Steele (1988) in the context of the MST and the greedy matching heuristic, respectively. Later, Rhee (1993b) investigated smoothness conditions for the minimal matching functional. She was the first to recognize that smoothness conditions yield isoperimetric inequalities which show that a Euclidean functional is close to its mean value. This insight is addressed more fully in Chapter 6.

Definition 3.4. (smoothness) A Euclidean functional L^p of order p is *smooth of order p* if there is a finite constant $C_3 := C_3(d,p)$ such that for all sets F, $G \subset [0,1]^d$ we have

$$(3.8) \qquad |L^p(F \cup G) - L^p(F)| \leq C_3(\mathrm{card}G)^{(d-p)/d}.$$

There are some simple consequences of smoothness (3.8), which we interpret as a Hölder continuity condition. First we notice that smoothness implies

$$|L^p(F) - L^p(G)| \leq 2C_3(\mathrm{card}(F \bigtriangleup G))^{(d-p)/d},$$

where $F \bigtriangleup G$ denotes the symmetric difference of the sets F and G. For point sets F and G in a general rectangle R, homogeneity gives

$$|L^p(F \cup G, R) - L^p(F, R)| \leq C_3(\mathrm{diam}R)^p(\mathrm{card}G)^{(d-p)/d}$$

for a possibly different choice of constant C_3. We will make frequent use of these smoothness inequalities.

The next result suggests that smoothness is rather ubiquitous. This result is not exhaustive and we will see that smoothness is a property common to many Euclidean functionals.

Lemma 3.5. When $0 < p \leq d$ the subadditive Euclidean functionals T^p, M^p, and S^p are smooth of order p. When $1 \leq p \leq d$ their superadditive canonical boundary versions T_B^p, M_B^p, and S_B^p are smooth of order p.

Proof. There is no single approach which yields smoothness (3.8) simultaneously for all three functionals and we therefore prove this on a case by case basis. Still there are some common ideas. For instance, letting L denote either of the functionals of Lemma 3.5, simple subadditivity (2.2) and the growth bounds (3.7) imply that for all sets F, $G \subset [0,1]^d$ we have $L^p(F \cup G) \leq L^p(F) + (C_1 + C_2 d^{p/2})(\mathrm{card}G)^{(d-p)/d} \leq L^p(F) + C(\mathrm{card}G)^{(d-p)/d}$. Thus, (3.8) follows once we show the reverse inequality

$$(3.9) \qquad L^p(F \cup G) \geq L^p(F) - C(\mathrm{card}G)^{(d-p)/d}.$$

To show (3.9) for the minimal spanning tree functional M^p, let T denote the graph of the minimal spanning tree on $F \cup G$. Remove the edges in T which contain a vertex in G. Since each vertex has bounded degree, say D, this generates a subgraph $T_1 \subset T$ which has at most $D \cdot \mathrm{card}G$ components. Choose one vertex from each component and form the minimal spanning tree T_2 on these vertices. Since the union of the trees T_1 and T_2 is a feasible spanning tree on F, it follows that

$$M^p(F) \leq \sum_{e \in T_1 \cup T_2} |e|^p \leq M^p(F \cup G) + C(D \cdot \mathrm{card}G)^{(d-p)/d}$$

by Lemma 3.3. Thus (3.8) holds for the MST functional M^p.

To see that (3.9) holds for the minimal matching functional S^p, we argue as follows. From the global matching on $F \cup G$, remove the edges which are incident to a vertex in G. This yields a collection $\mathcal{V}$ of unmatched vertices in F, where $\mathrm{card}\mathcal{V} \leq \mathrm{card}G$. By simple subadditivity of S^p we have

$$S^p(F) \leq S^p(F \cup G) + S^p(\mathcal{V}) \leq S^p(F \cup G) + C(\mathrm{card}G)^{(d-p)/d}$$

and thus S^p satisfies (3.9) and is smooth.

Finally, we show that (3.9) holds for the TSP functional T^p. When $p = 1$, (3.9) follows at once from monotonicity, that is from the relation $T(F) \leq T(F \cup \{x\})$ for all subsets $F \subset \mathbb{R}^d$ and singletons $x \in \mathbb{R}^d$. When $p > 1$ we need a more involved approach.

From the minimal tour on $F \cup G$, remove the edges which are incident to a vertex in G. This generates at most $\mathrm{card}G$ disconnected paths; letting $\mathcal{E}$ denote the collection of the endpoints of these paths we observe that $\mathrm{card}\mathcal{E} \leq 2\mathrm{card}G$. Consider $S^p(\mathcal{E})$, the length of the minimal matching on $\mathcal{E}$. This matching, together with the disconnected paths, generates a collection of tours $\{G_i\}_{i=1}^N$ on F, where N is random. From each tour G_i, $1 \leq i \leq N$, select an edge E_i which is generated by the minimal matching. Let e_i denote one endpoint of E_i. Let $\mathcal{E}' := \{e_i\}_{i=1}^N$ and consider $T^p(\mathcal{E}')$, the length of the minimal tour through $\mathcal{E}'$. The minimal tour consists of oriented edges $\{H_i\}_{i=1}^N$.

We now construct a tour through F by replacing the edges $\{H_i\}_{i=1}^N$ and $\{E_i\}_{i=1}^N$ according to the following simple rule. Observing that each edge $H \in \{H_i\}_{i=1}^N$ leads to one endpoint of an edge $E := E(H) \in \{E_i\}_{i=1}^N$, we replace the pair of edges H and E by the single edge joining the tail of H to the other endpoint of E. We may perform this replacement operation for all edges $H \in \{H_i\}_{i=1}^N$ at an extra cost of at most

$$C(S^p(\mathcal{E}) + T^p(\mathcal{E}')).$$

Moreover, this construction generates a feasible tour through F. We have thus shown that

$$T^p(F) \leq T^p(F \cup G) + C(S^p(\mathcal{E}) + T^p(\mathcal{E}')).$$

Since both $S^p(\mathcal{E})$ and $T^p(\mathcal{E}')$ are both bounded above by $C(\mathrm{card}\mathcal{E})^{(d-p)/d}$ and $\mathrm{card}\mathcal{E} \leq 2\mathrm{card}G$, the estimate (3.9) follows as desired.

We have thus shown that the functionals T^p, M^p, and S^p are smooth of order p. The proof that the boundary functionals T_B^p, M_B^p, and S_B^p are smooth of order p follows verbatim. This completes the proof of Lemma 3.5. $\square$

3.4. Closeness of a Functional to Its Canonical Boundary Functional

Superadditivity (3.3) and subadditivity (3.4) become especially useful when the boundary functional L_B^p is close to the standard functional L^p. Closeness of the two functionals can be measured in a deterministic (sup norm) sense or in a probabilistic sense. The first way of measuring closeness, given by the following definition, is especially valuable in the analysis of partitioning heuristics and large deviations, discussed in Chapters 5 and 6, respectively.

Definition 3.6. Say that L^p and L_B^p are *pointwise close* if for all subsets $F \subset [0,1]^d$ we have

$$(3.10) \qquad |L^p(F, [0,1]^d) - L_B^p(F, [0,1]^d)| = o\left((\text{card}F)^{(d-p)/d}\right).$$

Many functionals are pointwise close to the canonical boundary functional. The following lemma, which is not exhaustive, illustrates this.

Lemma 3.7. *The TSP, MST, and minimal matching functionals are pointwise close to their respective boundary functionals for $1 \leq p < d$ and in fact for all $F \subset [0,1]^d$ satisfy the estimate*

$$|L^p(F, [0,1]^d) - L_B^p(F, [0,1]^d)| \leq \begin{cases} C(\text{card}F)^{(d-p-1)/(d-1)}, & 1 \leq p < d-1, \\ C\log(\text{card}F), & p = d-1 \neq 1, \\ C, & d-1 < p < d, \\ & p = d-1 = 1. \end{cases}$$

Proof. We will sketch the proof for the minimal matching functional S^p. When $p = 1$ it clearly suffices to prove

$$(3.11) \qquad S(F, [0,1]^d) \leq S_B(F, [0,1]^d) + C(\text{card}F)^{(d-2)/(d-1)}.$$

Given $S_B(F, [0,1]^d)$, let $\mathcal{B}$ be the set of points where the edges in the graph which realizes $S_B(F, [0,1]^d)$ meet the boundary of $[0,1]^d$. Let $S(\mathcal{B})$ denote the length of a minimal matching on $\mathcal{B}$ with edges lying on $\partial[0,1]^d$. By simple subadditivity and Lemma 3.3 we have

$$S(F, [0,1]^d) \leq S_B(F, [0,1]^d) + S(\mathcal{B}) \leq S_B(F, [0,1]^d) + C(\text{card}\mathcal{B})^{(d-2)/(d-1)}$$

since $\mathcal{B}$ lies on the boundary, which has dimension $d - 1$. See Figure 3.1. Since $\text{card}\mathcal{B} \leq \text{card}F$, this proves (3.11).

Similar methods establish pointwise closeness for the MST and TSP functionals when $p = 1$. For more general p, $1 < p < d$, we require estimates for the sum of the lengths of the edges in $S_B^p(F)$ which meet the boundary. The following lemma is helpful.

Figure 3.1. Match the boundary points with the dotted edges to construct a feasible matching

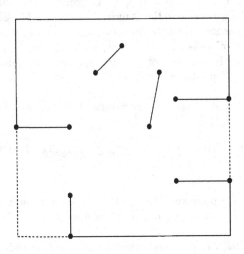

Lemma 3.8. *Let $F \subset [0,1]^d$, $\mathrm{card} F = n$, and consider the graph realizing the boundary minimal matching functional $S_B^p(F)$. Let $1 \le p \le d-1$. The sum of the pth powers of the lengths of the edges connecting vertices in F with the boundary of $[0,1]^d$ is bounded by $C(n^{(d-p-1)/(d-1)} \vee \log n)$. For $d-1 < p < d$ the sum is bounded by a constant C. The same estimates hold for the boundary MST functional M_B^p and the boundary TSP functional T_B^p.*

Proof of Lemma 3.8. We first prove the lemma for the TSP functional T^p, which is the more difficult case. The proof depends upon a dyadic subdivision of $[0,1]^d$. Let Q_0 be the cube of edge length $1/3$ and centered within $[0,1]^d$. Let Q_1 be the cube of edge length $2/3$, also centered within $[0,1]^d$. Partition $Q_1 - Q_0$ into subcubes of edge length $1/6$; it is easy to verify that the number of such subcubes is bounded by $C6^{d-1}$.

Continue with the subdivision recursively, so that at the jth stage we define cube Q_j of edge length $1 - 2(3 \cdot 2^j)^{-1}$ and partition $Q_j - Q_{j-1}$ into subcubes of edge length $(3 \cdot 2^j)^{-1}$. The number of such subcubes is at most $C3^{d-1}(2^j)^{d-1}$. Carry out this recursion until the kth stage, where k is the unique integer chosen so that

$$2^{(k-1)(d-1)} \le n < 2^{k(d-1)}.$$

This procedure produces nested cubes $Q_1 \subset Q_2 \subset ... \subset Q_k$. It produces a dyadic covering of the cube until the moat $[0,1]^d - Q_k$ has a width of the order $n^{-1/(d-1)}$. We use these properties to prove Lemma 3.8 as follows.

This dyadic subdivision partitions the largest cube Q_k into at most

$$\sum_{j=0}^{k} C3^{d-1}2^{j(d-1)} \le Cn$$

subcubes, each with an edge length equal to the distance between the subcube and the boundary of $[0,1]^d$. Furthermore, by partitioning each subcube of this partition into $2^{\ell d}$ congruent subcubes, where ℓ is the least integer satisfying $2^\ell \geq d^{1/2}$, we obtain a partition $\mathcal{P}$ of Q_k consisting of at most Cn subcubes with the property that the diameter of each subcube is less than the distance between it and the boundary.

Observe that in an optimal boundary tour on F each subcube Q in $\mathcal{P}$ contains at most two points in F which are rooted to the boundary. Indeed, were three or more points in $F \cap Q$ rooted to the boundary, then by minimality it would be more efficient to link two of these three points with an edge, since the diameter of the subcube is less than the distance to the boundary. See Figure 3.2.

Figure 3.2 Bounding the number of edges joined to the boundary: it is more efficient to insert edge AB

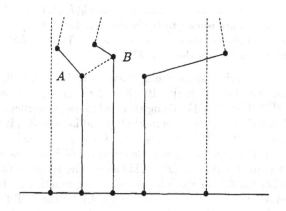

The sum of the pth powers of the lengths of the edges connecting vertices in $F \cap (Q_j - Q_{j-1})$ with the boundary is thus bounded by the product of the number of subcubes in $Q_j - Q_{j-1}$ and the pth power of the common diameter of the subcubes, namely

$$C3^{d-1}2^{j(d-1)} \cdot (3 \cdot 2^j)^{-p}.$$

Summing over all $1 \leq j \leq k$ gives a bound for the sum of the pth powers of the lengths of the edges connecting points in $F \cap Q_k$:

$$(3.12) \quad \sum_{j=1}^k C3^{d-1}2^{j(d-1)}(3 \cdot 2^j)^{-p} \leq \begin{cases} C(n^{(d-p-1)/(d-1)} \vee \log n), & 1 \leq p \leq d-1 \\ C, & d-1 < p < d. \end{cases}$$

The $\log n$ term is needed to cover the case $p = d-1$. The sum of the pth powers of the lengths of the edges connecting vertices in $F \cap ([0,1]^d - Q_k)$ with the boundary is at most the product of $n := \operatorname{card} F$ and the pth power of the width of the moat $[0,1]^d - Q_k$, i.e, at most

$$(3.13) \quad Cn \cdot n^{-p/(d-1)} = Cn^{(d-p-1)/(d-1)}.$$

Combining (3.12) and (3.13) establishes Lemma 3.8 for the TSP functional.

The proof for the analogous estimates involving the MST and minimal matching functionals is identical, save for the observation that there is *at most one vertex* in each subcube of $\mathcal{P}$ which is joined to the boundary. This concludes the proof of Lemma 3.8. $\square$

We now conclude the proof of Lemma 3.7 for the minimal matching functional S^p, $1 < p < d$. Recalling that $n = \text{card}F$, we need to show

$$(3.14) \qquad S^p(F, [0,1]^d) \leq \begin{cases} S_B^p(F, [0,1]^d) + Cn^{(d-p-1)/(d-1)}, & 1 \leq p < d-1 \\ S_B^p(F, [0,1]^d) + C\log n, & p = d-1 \\ S_B^p(F, [0,1]^d) + C, & d-1 < p < d. \end{cases}$$

Consider the minimal boundary matching which realizes $S_B^p(F, [0,1]^d)$ and let $F' \subset F$ denote those vertices which are rooted to the boundary. Let $\mathcal{B} \subset \partial[0,1]^d$ denote the set of points where the rooted edges meet the boundary. Thus $\text{card}F' = \text{card}\mathcal{B}$. Our goal is to construct a feasible matching on F'.

Construct the graph G which realizes the optimal matching $S^p(\mathcal{B}, [0,1]^d)$ and which has edges on the boundary of $[0,1]^d$. By (3.6) the edges in G have a total length of at most $C(n^{(d-p-1)/(d-1)} \vee 1)$. Using these edges we construct a natural pairing of points in F', which by the triangle inequality and Lemma 3.8, is achieved at a cost which is at most $C(n^{(d-p-1)/(d-1)} \vee \log n)$ for $1 < p \leq d-1$ and which is at most C for $d-1 < p < d$. This produces a feasible matching of F and therefore has a length which is clearly greater than $S^p(F)$. This shows the estimate (3.14) and concludes the proof of Lemma 3.7 for the minimal matching functional. The proof for the MST functional M^p is similar. To prove Lemma 3.7 for the TSP functional we may modify the proof of Lemma 3.10 below. This concludes the proof of Lemma 3.7. $\square$

The pointwise closeness of functionals (3.10) expresses an estimate which is usually more than sufficient for most approximation purposes. There is a second useful way to measure closeness, one which will be sufficient for finding asymptotics and rates of convergence of means.

Definition 3.9. (close in mean) Let L^p be a Euclidean functional and L_B^p the boundary functional, $1 \leq p < d$. L^p and L_B^p are *close in mean* if

$$(3.15) \qquad |EL^p(U_1, ..., U_n) - EL_B^p(U_1, ..., U_n)| = o(n^{(d-p)/d}).$$

Pointwise closeness clearly implies closeness in mean. However, the approximation error associated with closeness in mean is usually smaller than the corresponding approximation error associated with pointwise closeness. The following lemma shows that this is the case for our prototypical Euclidean functionals; later we will see that Lemma 3.10 holds for many other Euclidean functionals as well.

Lemma 3.10. *Let* $1 \leq p < d$. *The pth power weighted TSP, MST, and minimal matching functionals are close in mean to their respective boundary functionals and satisfy the approximation*

$$(3.16) \qquad |EL^p(U_1, ..., U_n) - EL_B^p(U_1, ..., U_n)| \leq M(d, p, n),$$

where $M(d, p, n) := Cn^{(d-p-1)/d}$ *for* $1 \leq p < d-1$, $M(d, p, n) := C \log n$ *for* $p = d - 1$, *and* $M(d, p, n) := C$ *for* $d - 1 < p < d$.

Proof. We will prove Lemma 3.10 for the TSP functional; the proofs for the MST and minimal matching functionals are similar. Since $T_B^p \leq T^p$, it suffices to show

$$(3.17) \qquad ET^p(U_1, ..., U_n,) \leq ET_B^p(U_1, ..., U_n) + M(d, p, n).$$

Let F denote one of the faces of $[0, 1]^d$. Letting $\mathcal{U}_F \subset \{U_1, ..., U_n\}$ be the set of points that are rooted to F by T_B^p, we first show that $E \text{card } \mathcal{U}_F \leq Cn^{(d-1)/d}$. For all $\epsilon > 0$ and $x \in F$, let $C(\epsilon, x)$ denote the cylinder in $[0, 1]^d$ determined by the ϵ disk in F centered at x. We now make the crucial observation that in the part of $C(\epsilon, x)$ which is at a distance greater than ϵ from F, there are at most two sample points which are joined to F by T_B^p. Were there three or more points, then two of these points could be joined with an edge, which would result in a cost savings, contradicting optimality (recall Figure 3.2). Since F can be covered with $O(\epsilon^{-(d-1)})$ disks of radius ϵ, we have the bound

$$E \text{card } \mathcal{U}_F \leq E \text{card}\{x \in (U_i)_{i \leq n} : d(x, F) \leq \epsilon\} + C\epsilon^{-(d-1)},$$

where $d(x, F)$ denotes the distance between the point x and the set F. The above is bounded by $n\epsilon + C\epsilon^{-(d-1)}$ and so putting $\epsilon = n^{-1/d}$ gives the desired estimate $E \text{card } \mathcal{U}_F \leq Cn^{(d-1)/d}$. If $\mathcal{U} \subset \{U_1, ..., U_n\}$ denotes the set of sample points which are rooted to any face of the boundary, then

$$(3.18) \qquad E \text{card } \mathcal{U} \leq Cn^{(d-1)/d}.$$

To prove (3.17), consider the boundary tour T which achieves $T_B^p(U_1, ..., U_n, [0, 1]^d)$. Let $\mathcal{B}$ denote the set of points where T meets the boundary; each point in $\mathcal{B}$ is thus the endpoint of some rooted path through some subset of $\{U_1, ..., U_n\}$. Let $S^p(\mathcal{B})$ denote the length of the minimal matching on $\mathcal{B}$ whose edges lie on $\partial[0, 1]^d$. Since $\mathcal{B}$ lies in a subset of dimension $d - 1$, it follows from the growth bounds (3.6) that $S^p(\mathcal{B}) \leq C((\text{card}\mathcal{B})^{(d-p-1)/(d-1)} \vee 1)$. By the estimate (3.18) and Jensen's inequality it follows that

$$(3.19) \qquad ES^p(\mathcal{B}) \leq C(n^{(d-p-1)/d} \vee 1).$$

The matching G which realizes $S^p(\mathcal{B})$ takes paths rooted to the boundary and forms a collection of closed tours $(T_i)_{i=1}^N$ on the sample $\{U_1, ..., U_n\}$, where N is random. From each tour T_i, $1 \leq i \leq N$, select one edge E_i, $E_i \in G$. Let e_i denote one endpoint of E_i. Let $\mathcal{E} := \{e_i\}_{i=1}^N$ and let $T^p(\mathcal{E})$ be the length of the minimal tour through $\mathcal{E}$ with edges on $\partial[0,1]^d$. The minimal tour consists of oriented edges, say $\{H_i\}_{1 \leq i \leq N}$. Note that as in (3.19)

$$(3.20) \qquad ET^p(\mathcal{E}) \leq C(n^{(d-p-1)/d} \vee 1).$$

We now produce a tour through $\{U_1, ..., U_n\} \cup \mathcal{B}$ by replacing the edges $\{H_i\}_{i=1}^N$ and $\{E_i\}_{i=1}^N$ according to the following natural rule. Noting that each edge $H \in \{H_i\}_{1 \leq i \leq N}$ leads to one endpoint of an edge $E := E(H) \in \{E_i\}_{i=1}^N$, replace H and E by the edge F joining the tail of H to the other endpoint of E. Thus F joins two points in $\mathcal{B}$. Performing this replacement operation for all edges H in $\{H_i\}_{1 \leq i \leq N}$ incurs a total cost of at most $C(S^p(\mathcal{B}) + T^p(\mathcal{E}))$. Moreover, this construction generates a tour T' through $\{U_1, ..., U_n\} \cup \mathcal{B}$.

We may generate a tour through $\{U_1, ..., U_n\}$ from the tour T' by replacing each edge F according to the following natural scheme: if F has endpoints f_1 and f_2 then consider the three edges in T' which are joined to either f_1 or f_2. The union of these edges is a path of the form a, f_1, f_2, b and we may replace the three edges with the *single* edge ab at an additional cost of at most

$$C(\|f_1 - f_2\|^p + \|a - f_1\|^p + \|b - f_2\|^p).$$

We have thus shown

$$(3.21) \qquad T^p(U_1, ..., U_n) \leq T^p_B(U_1, ..., U_n) + C(S^p(\mathcal{B}) + T^p(\mathcal{E}) + \Sigma^p(\mathcal{B})),$$

where $\Sigma^p(\mathcal{B})$ denotes the sum of the pth powers of the lengths of the edges in T which meet the boundary. A variation of Lemma 3.8 (choose k so that $2^{(k-1)d} \leq n < 2^{kd}$) shows that $E\Sigma^p(\mathcal{B}) \leq M(d, p, n)$. Taking expectations in (3.21) and applying (3.19)-(3.20) gives the estimate (3.17), as desired. This concludes the proof of Lemma 3.10. $\square$

There is a second way to prove Lemma 3.10 which is similar to the proof of Lemma 3.8: we simply use a dyadic covering of the cube until the moat has width of the order $n^{-1/d}$. See Redmond and Yukich (1994) for details.

Notes and References

1. Condition (3.16) suggests that the functionals L^p and L_B^p are nearly additive, thus prompting Redmond and Yukich (1994) to use the appellation "quasi-additive".

2. The random version of the TSP received attention prior to the celebrated paper of Beardwood, Halton, and Hammersley (1959). We mention the work of Mahalanobis (1940), Jessen (1942), Marks (1948), and Few (1955).

3. We will not be concerned with the numerical value of the constant C_2 in (3.6). There has been nonetheless considerable work in this area. When L is the TSP functional, $p = 1$, and R is the unit square, for example, Few (1962) showed that C_2 can be taken to be $(4/3)^{(1/4)} + \epsilon$ for all sets F having cardinality larger than $N(\epsilon)$.

4. The space filling curve heuristic is described at length in Steele (1997). We have followed parts of his exposition very closely.

5. We expect that the $\log n$ term in the estimate (3.16) can be removed and that (3.16) can be improved to

$$|EL^p(U_1, ..., U_n) - EL_B^p(U_1, ..., U_n)| \leq C(n^{(d-p-1)/d} \vee 1).$$

This is the case if L is the MST functional (Redmond and Yukich (1996)).

4. ASYMPTOTICS FOR EUCLIDEAN FUNCTIONALS:

THE UNIFORM CASE

The goal of this chapter is to provide a simple and natural approach to finding the asymptotics of the lengths of graphs associated with problems in geometric probability. This includes the classic problems described in the previous chapters as well as others which are less widely known. The structural properties developed in Chapter 3 facilitate our approach.

Much of this chapter and indeed the entire monograph is inspired by the pioneering paper of Beardwood, Halton, and Hammersley (1959) who showed that the TSP functional T satisfies

$$(4.1) \qquad \lim_{n \to \infty} T(U_1, ..., U_n)/n^{(d-1)/d} = \alpha(T, d) \quad a.s.$$

Here and throughout $U_1, ..., U_n$ are i.i.d. with the uniform distribution on $[0,1]^d$ and $\alpha(T, d)$ is a positive constant depending only on the dimension d. Beardwood, Halton, and Hammersley actually proved much more and considered the behavior of the TSP functional on sequences of random variables more general than the uniform random variables; this generalization, while important, does not concern us here and we will return to it in Chapter 7.

The limit (4.1) is easily explained on intuitive grounds. Indeed, one expects that the average length of a typical edge in the tour on $U_1, ..., U_n$ would be of the order $n^{-1/d}$, the distance between a typical point and its nearest neighbor. Since there are n edges, one anticipates an average growth rate of $n^{(d-1)/d}$. The limit (4.1) makes this precise. In this chapter we show that (4.1) holds for general subadditive Euclidean functionals.

Borovkov (1962) was apparently the first to use (4.1) in the study of algorithms for the Euclidean TSP, an NP-complete problem (cf. Papadimitriou, 1978b). He used (4.1) to show the existence of a feasible tour whose length is a.s. within a small multiple of the minimal tour length.

Karp (1976, 1977) put (4.1) to striking use in the analysis of partitioning algorithms approximating the shortest tour. As mentioned earlier, Karp showed that in the stochastic setting we may use (4.1) to exhibit a polynomial time algorithm such that the algorithm a.s. provides a solution that is within a factor of $1 + \epsilon$ of the length of the minimal tour. Karp's algorithm involves a partitioning heuristic which consists of tying together $O(n/\log n)$ optimal tours on $O(n/\log n)$ subsquares to obtain a grand tour which suitably approximates the optimal tour. Rephrasing Karp's theorem, one often says that the partitioning heuristic is with probability one ϵ-optimal. Karp's results did much to stimulate interest in stochastic versions of combinatorial optimization problems. We will return to Karp's partitioning heuristic in Chapter 5.

The limit (4.1) and its generalizations also play an important role in the probabilistic evaluation of the performance of heuristic algorithms for vehicle routing problems. Haimovich and Rinnooy Kan (1985) use (4.1) to analyze heuristics for the capacitated vehicle routing problem. For a survey of these algorithms we refer to Karp and Steele (1985) and Haimovich et al. (1988).

In the sequel we will need to prove limit results in the sense of *complete convergence*. We recall that a sequence of random variables X_n, $n \geq 1$, converges completely (c.c.) to a constant C if and only if for all $\epsilon > 0$ we have

$$\sum_{n=1}^{\infty} P\{|X_n - C| > \epsilon\} < \infty.$$

The main attraction of complete convergence is not that it strengthens a.s. convergence, but that it provides convergence results for the two distinctly different ways to interpret the dependence of the functionals $L^p(X_1, ..., X_n)$ and $L^p(X_1, ..., X_n, X_{n+1})$. The dependence of these functionals on one another has not been made explicit. Given the functional $L^p(X_1, ..., X_n)$, one can increment the number of existing sample points by one to get the new functional $L^p(X_1, ..., X_n, X_{n+1})$; this is the so-called *incrementing model of problem generation*. However, one can also consider the functional which is based on a completely new sample of points $\{X'_1, ..., X'_n, X'_{n+1}\}$ to get the new functional $L^p(X'_1, ..., X'_n, X'_{n+1})$. This second method is the *independent model of problem generation*. The difference between the limit theory for the two models is analogous to the difference between the limit theory of sequences and triangular arrays of random variables. A.s. limit results for the independent model imply a.s. limits for the incrementing model, but without extra assumptions on L^p, the converse is false in general.

To prove a.s. limits for both models of problem generation we will show the complete convergence of $L^p(X_1, ..., X_n)/n^{(d-p)/d}$. Notice that the "hard" half of the Borel-Cantelli lemma shows that c.c. results are necessary if one is to obtain a.s. asymptotics in the context of the independent model. Weide (1978) was the first to recognize the need for complete convergence results in the probabilistic analysis of algorithms.

Thus, given an optimization problem in the form of a smooth subadditive Euclidean functional L^p, we henceforth will not specify the underlying model of problem generation since complete convergence handles both models alike.

4.1. Limit Theorems for Euclidean Functionals L^p, $1 \leq p < d$

When $p = 1$ the limiting a.s. behavior of the TSP functional T and its cousins M and S is relatively well understood. This is due largely to the seminal work of Steele (1981a, 1988, 1990b). Steele (1981a) showed that the asymptotics (4.1) for the TSP could be generalized to a wide class of optimization functionals. He showed that if a subadditive Euclidean functional L is *monotone* in the sense that $L(F) \leq L(F \cup x)$ for sets $F \subset \mathbb{R}^d$ and singletons $x \in \mathbb{R}^d$ then L enjoys the asymptotics

$$\text{(4.2)} \qquad \lim_{n \to \infty} L(U_1, ..., U_n)/n^{(d-1)/d} = \alpha(L, d) \quad a.s.,$$

where $\alpha(L, d)$ is a positive constant depending only on the functional L and the dimension d. Since the TSP is monotone, (4.2) obviously gives asymptotics for the TSP functional as a special case.

However, many Euclidean functionals, including the minimal spanning tree and minimal matching functionals, are not monotone. Inserting points may reduce the total edge length in the minimal spanning tree and minimal matching graphs. This source of bedevilment has been an obstacle to the development of asymptotics.

Rhee (1993b) overcame this obstacle. She recognized that (4.2) continues to hold whenever the Euclidean functional L is *smooth* in the sense of (3.8), namely whenever for all F, $G \subset [0, 1]^d$ we have

$$|L^p(F \cup G) - L^p(F)| \leq C_3 (\text{card} G)^{(d-p)/d}.$$

We recall from Lemma 3.5 that the TSP, MST, and minimal matching functionals are smooth and in the sequel we will see that many other Euclidean functionals are smooth. By using smoothness as a substitute for monotonicity, Rhee captures the a.s. asymptotics for the TSP, MST, and minimal matching functionals.

In the spirit of Rhee (1993b) we now formulate a general asymptotic result lying at the heart of our subject.

Theorem 4.1. *(basic limit theorem for Euclidean functionals L^p, $1 \leq p < d$) If L_B^p is a smooth superadditive Euclidean functional of order p on $\mathbb{R}^d$, $1 \leq p < d$, then*

$$\text{(4.3)} \qquad \lim_{n \to \infty} L_B^p(U_1, ..., U_n)/n^{(d-p)/d} = \alpha(L_B^p, d) \quad c.c.,$$

where $\alpha(L_B^p, d)$ is a positive constant. If L^p is a Euclidean functional of order p on $\mathbb{R}^d$, $1 \leq p < d$, which is pointwise close to L_B^p, then

$$\text{(4.4)} \qquad \lim_{n \to \infty} L^p(U_1, ..., U_n)/n^{(d-p)/d} = \alpha(L_B^p, d) \quad c.c.$$

Remark. Since $L^p(U_1, ..., U_n) \le C_3 n^{(d-p)/d}$ always holds for smooth L, it follows by the bounded convergence theorem that (4.3) and (4.4) also hold for the means of L_B^p and L^p.

Since the optimization functionals T^p, M^p, and S^p are smooth subadditive Euclidean functionals (Lemma 3.5) which are pointwise close to their respective boundary versions (Lemma 3.7), we immediately deduce our first asymptotic result for optimization functionals on random samples:

Corollary 4.2. *Let* $1 \le p < d$. *If* L^p *denotes either the TSP, MST, or minimal matching functional of order* p, *then*

$$\lim_{n \to \infty} L^p(U_1, ..., U_n)/n^{(d-p)/d} = \alpha(L_B^p, d) \;\; c.c.$$

Later, we will see that the basic limit Theorem 4.1 admits an extension which allows us to replace the uniform random variables by arbitrary i.i.d. random variables X_i, $i \ge 1$. The only additional assumption required for this extension is closeness in mean (3.15) of L^p and L_B^p. See Theorems 7.1 and 7.5.

Proof of Theorem 4.1. (Sketch) We only prove a mean version of (4.3), namely

$$(4.5) \qquad\qquad \lim_{n \to \infty} EL_B^p(U_1, ..., U_n)/n^{(d-p)/d} = \alpha(L_B^p, d).$$

Later, isoperimetric methods will show that the mean version is equivalent to the c.c. version (see Corollary 6.4 of Chapter 6).

To prove (4.5), fix $1 \le p < d$ and set $\phi(n) := EL_B^p(U_1, ..., U_n)$. Observe that the number of points from the sample $(U_1, ..., U_n)$ which fall in a given subcube of $[0,1]^d$ of volume m^{-d} is a binomial random variable $B(n, m^{-d})$ with parameters n and m^{-d}. It follows from the superadditivity of L_B^p, homogeneity (3.2), smoothness (3.8), and Jensen's inequality that

$$\phi(n) \ge m^{-p} \sum_{i \le m^d} \phi(B(n, m^{-d}))$$

$$\ge m^{-p} \sum_{i \le m^d} \left(\phi(nm^{-d}) - C_3 E(|B(n, m^{-d}) - nm^{-d}|^{(d-p)/d}) \right)$$

$$\ge m^{-p} \sum_{i \le m^d} \left(\phi(nm^{-d}) - C_3 (nm^{-d})^{(d-p)/2d} \right).$$

Simplifying, we get

$$\phi(n) \ge m^{d-p} \phi(nm^{-d}) - C_3 m^{(d-p)/2} n^{(d-p)/2d}.$$

Dividing by $n^{(d-p)/d}$ and replacing n by nm^d yields the homogenized relation

(4.6)
$$\frac{\phi(nm^d)}{(nm^d)^{(d-p)/d}} \geq \frac{\phi(n)}{n^{(d-p)/d}} - \frac{C_3}{n^{(d-p)/2d}}.$$

Set $\alpha := \alpha(L_B^p, d) := \limsup_{n \to \infty} \phi(n)/n^{(d-p)/d}$ and note that $\alpha \leq C_3$ by the assumed smoothness. For all $\epsilon > 0$, choose n_o such that for all $n \geq n_o$ we have $C_3/n^{(d-p)/2d} \leq \epsilon$ and $\phi(n_o)/n_o^{(d-p)/d} \geq \alpha - \epsilon$. Thus, for all $m = 1, 2, \dots$ it follows that

$$\phi(n_o m^d)/(n_o m^d)^{(d-p)/d} \geq \alpha - 2\epsilon.$$

To now obtain (4.5) we use the smoothness of L and a simple interpolation argument. For an arbitrary integer $k \geq 1$ find the unique integer m such that

$$n_o m^d < k \leq n_o (m+1)^d.$$

Then $|n_o m^d - k| \leq C n_o m^{d-1}$ and by smoothness (3.8) we therefore obtain

$$\frac{\phi(k)}{k^{(d-p)/d}} \geq \frac{\phi(n_o m^d)}{(n_o(m+1)^d)^{(d-p)/d}} - \frac{(Cn_o m^{d-1})^{(d-p)/d}}{(m+1)^{d-p} \, n_o^{(d-p)/d}}$$

$$\geq (\alpha - 2\epsilon)(\frac{m}{m+1})^{d-p} - \frac{(Cn_o m^{d-1})^{(d-p)/d}}{(m+1)^{d-p} \, n_o^{(d-p)/d}}.$$

Since the last term in the above goes to zero as m goes to infinity, it follows that

(4.7)
$$\liminf_{k \to \infty} \phi(k)/k^{(d-p)/d} \geq \alpha - 2\epsilon.$$

Now let ϵ tend to zero to see that the liminf and the limsup of the sequence $\phi(k)/k^{(d-p)/d}$, $k \geq 1$, coincide, that is

$$\lim_{k \to \infty} \phi(k)/k^{(d-p)/d} = \alpha.$$

We have thus shown

$$\lim_{n \to \infty} EL_B^p(U_1, \dots, U_n)/n^{(d-p)/d} = \alpha$$

as desired. We will see in Section 4.4 that α is positive. This completes the proof of (4.5).

The limit (4.4) is automatic and the proof of Theorem 4.1 is complete. □

The above proof takes advantage of the self-similarity properties of the Euclidean functional L^p. The proof is pleasantly simple and self-contained. Notice that the proof breaks down when $p \geq d$; for these values of p we will need an approach which is more delicate and which is discussed in the next section.

If L^p, $1 \leq p < d$, is a smooth Euclidean functional which satisfies the subadditivity condition

$$(4.8) \qquad L^p(F, [0,1]^d) \leq \sum_{i=1}^{m^d} L^p(F \cap Q_i, Q_i) + C_1 m^{d-p},$$

where Q_i, $1 \leq i \leq m^d$, denotes a partition of $[0,1]^d$ into m^d subcubes of edge length m^{-1}, then (4.4) follows. To see this, follow the proof of (4.5) verbatim and notice that (4.6) becomes

$$(4.9) \qquad \frac{\phi(nm^d)}{(nm^d)^{(d-p)/d}} \leq \frac{\phi(n)}{n^{(d-p)/d}} + \frac{C_3}{n^{(d-p)/d}} + \frac{C_1 d^{1/2}}{n^{(d-p)/d}},$$

where $\phi(n) := EL(U_1, ..., U_n)$. The proof of (4.4) now follows the proof of (4.5).

We will henceforth not need the subadditivity condition (4.8) and instead we will rely upon the simpler subadditivity (3.4), which, as noted previously, implies (4.8) when m is a power of 2^d.

4.2. Limit Theorems for Euclidean Functionals L^p, $p \geq d$

The previous section developed a limit theorem for Euclidean functionals L^p when $1 \leq p < d$. This section discusses the more delicate case $p \geq d$. For many Euclidean functionals L^p the dth power of the length of a typical edge in the graph associated with $L^p(U_1, ..., U_n)$ is of the order n^{-1}. When there are $O(n)$ edges in the graph we would expect that $L^d(U_1, ..., U_n)$ would behave like a constant for large n. The main point of this section is to show that this is indeed the case. This is less straightforward than might first appear.

For $p \geq d$ the methods of Section 4.1 break down since they introduce nonnegligible constant terms in both the superadditive and subadditive relations (4.6) and (4.9) respectively. For these critical values of p, a more delicate approach is needed. By considering the probability theory of infinite trees, Aldous and Steele (1992) obtain L^2 asymptotics for the MST functional $M^d(U_1, ..., U_n)$. The Aldous and Steele approach spawned a number of interesting conjectures on infinite trees and their methods may possibly be useful in the context of the TSP and minimal matching functionals. We refer to Steele (1997) for a complete treatment.

In what follows we will continue to view optimization problems L as superadditive Euclidean functionals on the product space $\mathcal{F} \times \mathcal{R}(d)$. This approach delivers asymptotics for L^p in the critical case $p \geq d$ as well as in the case $1 \leq p < d$.

The following limit theorem provides asymptotics for the mean of Euclidean functionals L^p, $p \geq d$, over a Poisson number N of points. Here $N := N(n)$ is an independent Poisson random variable with parameter n. More about the constants $\alpha(L_B^p, d)$ will follow in section 4.4.

Theorem 4.3. *(basic limit theorem for Euclidean functionals L^p, $p \geq d$) Let L_B^p be a superadditive Euclidean functional of order p on $\mathbb{R}^d$, where $p \geq d \geq 2$. If $EL_B^p(U_1, ..., U_n) \leq Cn^{(d-p)/d}$, then*

$$(4.10) \qquad \lim_{n \to \infty} n^{(p-d)/d} EL_B^p(U_1, ..., U_N) = \alpha(L_B^p, d)$$

where $\alpha(L_B^p, d)$ is a positive constant. If L^p is close in mean to L_B^p in the sense that

$$(4.11) \qquad E|L^p(U_1, ..., U_N) - L_B^p(U_1, ..., U_N)| = o(n^{(d-p)/d}),$$

then

$$(4.12) \qquad \lim_{n \to \infty} n^{(p-d)/d} EL^p(U_1, ..., U_N) = \alpha(L_B^p, d).$$

To obtain asymptotics for the de-Poissonized versions of (4.10) and (4.12) we would normally appeal to smoothness (3.8). However, this smoothness is not sufficiently strong and we will instead consider a modified smoothness condition.

Definition 4.4. *(smooth in mean) For all $p > 0$ and $d \geq 2$ a Euclidean functional L^p is smooth in mean if there exists a constant $\gamma < 1/2$ such that for all $n \geq 1$ and $0 \leq k \leq n/2$ we have*

$$(4.13) \qquad E|L^p(U_1, ..., U_n) - L^p(U_1, ..., U_{n \pm k})| \leq Ckn^{-p/d+\gamma}.$$

The next result is a de-Poissonized version of Theorem 4.3.

Theorem 4.5. *Let L_B^p be a superadditive Euclidean functional of order p on $\mathbb{R}^d$, where $p \geq d \geq 2$. Assume that $EL_B^p(U_1, ..., U_n) \leq Cn^{(d-p)/d}$. If L_B^p is close in mean (4.11) to L^p and smooth in mean (4.13) then*

$$(4.14) \qquad \lim_{n \to \infty} n^{(p-d)/d} EL_B^p(U_1, ..., U_n) = \alpha(L_B^p, d)$$

and

$$(4.15) \qquad \lim_{n \to \infty} n^{(p-d)/d} EL^p(U_1, ..., U_n) = \alpha(L_B^p, d).$$

In the remainder of this section we consider the proofs of Theorems 4.3 and 4.5.

Proof of Theorem 4.3. Let $N(\lambda)$ be a Poisson random variable with parameter $\lambda > 0$ and which is independent of the sequence U_i, $i \geq 1$. Set

$$\phi(\lambda) := EL_B^p(U_1, ..., U_{N(\lambda)}).$$

The proof of (4.10) centers around two functional inequalities for a scaled version of ϕ defined by

$$h(\lambda) := \phi(\lambda)/\lambda^{(d-p)/d}.$$

Notice first that $\sup_\lambda h(\lambda) \leq C$. To see this, note that by the assumed boundedness of $EL^p_B(U_1, ..., U_n)$, the assumed independence of N and U_i, $i \geq 1$, and Fubini's theorem we have

$$EL^p_B(U_1, ..., U_{N(\lambda)}) \leq CE\left((N(\lambda))^{(d-p)/d} \cdot 1_{\{N(\lambda)>0\}}\right).$$

The right side of the above is bounded by

$$CE\left((N(\lambda))^{(d-p)/d} \cdot 1_{\{0<N(\lambda)<\lambda/2\}}\right) + CE\left((N(\lambda))^{(d-p)/d} \cdot 1_{\{\lambda/2\leq N(\lambda)<\infty\}}\right)$$
$$\leq C\lambda^{(d-p)/d}$$

using exponential bounds for $P\{0 < N(\lambda) < \lambda/2\}$.

To derive our first inequality, observe that the superadditivity of L^p_B implies for all $0 < \delta < 1$

$$L^p_B(U_1, ..., U_{N(\lambda)}) \geq L^p_B\left(\{U_1, ..., U_{N(\lambda)}\} \cap [0, 1-\delta]^d, [0, 1-\delta]^d\right).$$

Taking expectations and scaling gives

$$\phi(\lambda) \geq (1-\delta)^p \phi(\lambda(1-\delta)^d).$$

Dividing by $\lambda^{(d-p)/d}$ we obtain

$$h(\lambda) \geq (1-\delta)^d h(\lambda(1-\delta)^d),$$

where $\lambda > 0$ and $0 < \delta < 1$. Using $\sup_\lambda h(\lambda) \leq C$, we easily obtain our first inequality for h:

(4.16) $$h(\lambda) \geq h(\lambda(1-\delta)^d) - \delta C.$$

To derive a second functional relationship for h, partition $[0, 1]^d$ into m^d disjoint subcubes $Q_1, ..., Q_{m^d}$ of edge length m^{-1}. Superadditivity of L^p_B gives

$$L^p_B(U_1, ..., U_{N(\lambda)}) \geq \sum_{i=1}^{m^d} L^p_B\left(\{U_1, ..., U_{N(\lambda)}\} \cap Q_i, Q_i\right).$$

Taking expectations and scaling yields

$$\phi(\lambda) \geq m^{d-p} \phi(\lambda m^{-d}).$$

Thus for all $\lambda > 0$ and all positive integers m we get

$$\phi(m^d \lambda) \geq m^{d-p} \phi(\lambda).$$

Dividing by $\lambda^{(d-p)/d}m^{d-p}$ we obtain our second inequality valid for all $m \in \mathbb{N}$ and all $\lambda > 0$:

$$(4.17) \qquad\qquad\qquad h(m^d\lambda) \geq h(\lambda).$$

We now combine the functional relations (4.16) and (4.17) to deduce (4.10). Let $\limsup_{\lambda\to\infty} h(\lambda) := \beta$ and note that β is finite. For all $\epsilon > 0$ we may find $\lambda_o := \lambda_o(\epsilon)$ such that $h(\lambda_o) \geq \beta - \epsilon$. By (4.17) we thus have for all positive integers m

$$h(m^d\lambda_o) \geq \beta - \epsilon.$$

It remains to show that $\liminf_{\lambda\to\infty} h(\lambda) = \beta$. To do this, we examine how h fluctuates in the interval $[m^d\lambda_o, (m+1)^d\lambda_o)$. With ϵ fixed as above and $m \geq \epsilon^{-1}$, consider $t \in \mathbb{R}^+$ such that

$$m^d\lambda_o < t \leq (m+1)^d\lambda_o.$$

The relation (4.16) gives

$$h(t) \geq h(t(1-\delta)^d) - \delta C$$

for all $0 < \delta < 1$. Set $\delta := 1 - (\frac{m^d\lambda_o}{t})^{1/d}$. Note that since $t \leq m^d\lambda_o + Cm^{d-1}\lambda_o$ we obtain the estimate

$$\delta \leq 1 - (\frac{m}{m+C})^{1/d} \leq 1 - \frac{m}{m+C} \leq \frac{C}{m}.$$

We thus obtain

$$h(t) \geq h(m^d\lambda_o) - \delta C \geq h(m^d\lambda_o) - \frac{C}{m}.$$

Thus for all t between $m^d\lambda_o$ and $(m+1)^d\lambda_o$ we deduce from (4.17)

$$h(t) \geq h(m^d\lambda_o) - C\epsilon \geq \beta - \epsilon - C\epsilon.$$

Thus, $\liminf_{t\to\infty} h(t) \geq \beta - \epsilon - C\epsilon$. Let ϵ tend to zero and set $\beta := \alpha(L_B^p, d)$ to deduce (4.10). We will see in Section 4.4 that $\alpha(L_B^p, d)$ is positive. The limit (4.12) is immediate and the proof of Theorem 4.3 is complete. $\square$

We now consider the proof of Theorem 4.5.

Proof of Theorem 4.5. We need to show that (4.10) implies the de-Poissonized limit (4.14). Let A denote the event $\{|N - n| \leq C(n\log n)^{1/2}\}$ and for all $k \in \mathbb{N}$ let $L(k) := L(U_1, ..., U_k)$. Write the decomposition

$$n^{(p-d)/d}|EL_B^p(N) - EL_B^p(n)|$$
$$\leq n^{(p-d)/d}E(|L_B^p(N) - L_B^p(n)| \cdot 1_{A^c}) + n^{(p-d)/d}E(|L_B^p(N) - L_B^p(n)| \cdot 1_A).$$

For C large, the first term converges to zero as n tends to infinity by the assumed boundedness of $EL_B^p(n)$ as well as by the exponential tails for $N - n$. The second term is handled by a conditioning argument and by independence equals

$$= n^{(p-d)/d} \sum_{|k| \leq C(n \log n)^{1/2}} E\left(|L_B^p(n+k) - L_B^p(n)|\right) P\{N = n+k\}$$

$$\leq C n^{(p-d)/d} \sum_{|k| \leq C(n \log n)^{1/2}} k n^{-p/d+\gamma} P\{N = n+k\}$$

$$\leq C n^{-1+\gamma} \sum_{k=-n}^{\infty} k P\{N - n = k\}$$

$$\leq C n^{-1+\gamma} E|N - n|$$

$$= o(1).$$

Thus the second term goes to zero as well, showing that (4.10) implies (4.14). The proof that (4.12) implies (4.15) is similar. $\square$

4.3. Applications to Problems in Combinatorial Optimization

Since the edges in the graphs of the minimal spanning tree and the shortest tour on n uniform random variables have an average length of the order $n^{-1/d}$, we would expect that the sum of the dth powers of the lengths of these edges would behave like a constant. The main point of this section is to justify these heuristics with the aid of the limit Theorems 4.3 and 4.5. This section is not essential to the sequel and may be skipped without loss of continuity.

Throughout, we adhere to the convention that U_i, $i \geq 1$, denotes an i.i.d. sequence of uniform random variables on $[0,1]^d$ and $N := N(n)$ denotes an independent Poisson random variable with parameter n.

Theorem 4.6. (*asymptotics for the power-weighted MST*) *For all $p \geq d \geq 2$ we have*

$$(4.18) \qquad \lim_{n \to \infty} n^{(p-d)/d} EM^p(U_1, ..., U_n) = \alpha(M_B^p, d).$$

Theorem 4.7. (*asymptotics for the power-weighted TSP*) *For all $d \geq 2$*

$$(4.19) \qquad \lim_{n \to \infty} ET^d(U_1, ..., U_n) = \alpha(T_B^d, d).$$

The basic limit Theorems 4.3 and 4.5 could also be used to capture the asymptotics for the minimal matching functional and related optimization problems. However we will not pursue this and instead we sketch the proof of Theorem 4.6; the somewhat involved proof of Theorem 4.7 appears in Yukich (1995a).

The interest of Theorems 4.6 and 4.7 derives from the fact that they hold for the critical case $p = d$, where the usual geometric subadditivity methods break down. As early as 1968 Gilbert and Pollak (1968) proved that the MST functional $M^2(F)$ is uniformly bounded over $F \subset [0,1]^2$, but they didn't address the issue of convergence. When $p = d$, Bland and Steele conjectured that (4.18) and (4.19) hold respectively.

When $p = d$, Aldous and Steele (1992) considered the probabilistic theory of infinite trees in $\mathbb{R}^d$ to obtain an L^2 version of (4.18), thus settling Bland's conjecture. Aldous and Steele use the "objective method" (cf. Steele, 1997) to prove an L^2 version of (4.18). They were motivated to use this approach since at the time it appeared that (4.18) was not within reach of the usual subadditivity methods. It is not clear whether the objective method can be used for the minimal matching and TSP functionals. Later, Yukich (1995a) used boundary functionals to establish (4.18) and (4.19), thus settling Steele's conjecture.

In order to prove Theorem 4.6 we will call upon the following lemma which shows that the MST functional satisfies the smooth in mean condition (4.13). We continue to write $M(n)$ for $M(U_1, ..., U_n)$.

Lemma 4.8. *(smoothness for the MST functional). For all $p > 0$ and $d \geq 2$ there is a $C := C(d, p)$ such that for all $n \geq 1$ and $0 \leq k \leq n/2$ we have*

$$(4.20) \qquad |M_B^p(n) - M_B^p(n+k)| \leq Ck(\log n/n)^{p/d}$$

with high probability. Moreover, for all $0 \leq k \leq n/2$

$$(4.21) \qquad |M_B^p(n) - M_B^p(n-k)| \leq Ck(\log n/n)^{p/d}$$

with high probability.

The precise meaning of the "high probability" statement (4.20) is as follows. For any prescribed $\beta > 0$ we can find $C := C(\beta)$ such that for any $0 \leq k \leq n/2$

$$P\{|M_B^p(n) - M_B^p(n+k)| > Ck(\log n/n)^{p/d}\} = O(n^{-\beta}).$$

A similar meaning is attached to the statement (4.21) as well as related high probability statements in this monograph.

Proof of Lemma 4.8. Notice that given the graph of the minimal spanning tree on $(U_1, ..., U_n)$, we can construct a feasible spanning tree on $(U_1, ..., U_n, U_{n+1})$ by inserting the edge of minimal length between U_{n+1} and the sample $(U_1, ..., U_n)$. Thus

$$M_B^p(n+1) \leq M_B^p(n) + d^p(U_{n+1}, (U_i)_{i=1}^n),$$

where $d(x, F)$ denotes the Euclidean distance between the point x and the set F. In general, for $k \geq 1$ we have

$$M_B^p(n+k) \leq M_B^p(n) + \sum_{j=n+1}^{n+k} d^p(U_j, (U_i)_{i=1}^n).$$

For all $j \geq n+1$ we have by standard arguments that $d(U_j, (U_i)_{i=1}^n) \leq C(\log n/n)^{1/d}$ with high probability. We obtain for all $k \geq 0$ the high probability estimate

$$M_B^p(n + k) \leq M_B^p(n) + Ck(\log n/n)^{p/d}.$$

To complete the proof of (4.20) we need to show the reverse high probability inequality

$$M_B^p(n) \leq M_B^p(n + k) + Ck(\log n/n)^{p/d}.$$

We will first show the high probability inequality

(4.22) $$M_B^p(n) \leq M_B^p(n + 1) + C(\log n/n)^{p/d}$$

and then iterate. Let $(N_j)_{j=1}^{M(d)} \subset \{U_1, ..., U_n\}$ denote the vertices which are linked to U_{n+1} by the minimal spanning tree on $U_1, ..., U_n, U_{n+1}$. $M(d)$ is finite since vertices in minimal spanning trees have bounded degree (see e.g. Melzak (1973)). With high probability there is a sample point, which without loss of generality we label U_1, such that U_{n+1} is within $C(\log n/n)^{1/d}$ of U_1. Replace all $M(d)$ edges E_i, $1 \leq i \leq M(d)$, having U_{n+1} as a vertex with edges E_i' leading to U_1 instead. See Figure 4.1.

Figure 4.1.   Constructing a feasible spanning tree: replace edges E_1 and E_2 with the dashed edges

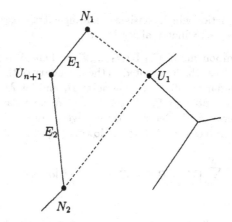

For each $1 \leq i \leq M(d)$, this may be achieved at a cost of at most

$$(|E_i| + C(\log n/n)^{1/d})^p \leq C(\log n/n)^{p/d}$$

since with high probability $|E_i| \leq C(\log n/n)^{1/d}$, a fact which we leave as an exercise. The resulting graph gives a feasible spanning tree on the pruned set $U_1, ..., U_n$ and shows the high probability bound

$$M_B^p(n) \leq M_B^p(n+1) + C(\log n/n)^{p/d},$$

which is precisely (4.22). Iterating gives (4.20). The proof of (4.21) is similar. This completes the proof of Lemma 4.8. $\square$

We are now ready for the

Proof of Theorem 4.6.

It is straightforward to show that the MST functional M_B^p is superadditive and satisfies $E M_B^p(U_1, ..., U_n) \leq C n^{(d-p)/d}$. By Theorem 4.3 we have

$$\lim_{n \to \infty} n^{(p-d)/d} E M_B^p(U_1, ..., U_N) = \alpha(L_B^p, d).$$

To show (4.18) we will show that M_B^p satisfies the conditions of Theorem 4.5. By Lemma 4.8 we already know that smoothness in mean (4.13) is satisfied. It remains to show closeness in mean, i.e., show for all $p \geq d \geq 2$ that

$$(4.23) \qquad\qquad E|M^p(n) - M_B^p(n)| = o(n^{(d-p)/d}).$$

We rely upon a construction which consists of adding extra edges to the components in the graph $G_B := G_B^p(n)$ which realizes $M_B^p(n)$.

Enumerate the components of G_B by $T_1, ..., T_N$, where N is random and where each T_i represents a tree which is rooted to the boundary of the unit cube. Let the tree T_i meet the boundary of $[0,1]^d$ at the point B_i and let M_i denote the unique sample point which is rooted to B_i, $1 \leq i \leq N$. We now want to show that the sum of the pth powers of the lengths of the edges connected to the boundary is $o(n^{(d-p)/d})$. We will in fact establish the following high probability bound

$$(4.24) \qquad\qquad \sum_{i=1}^{N} |M_i - B_i|^p \leq C n^{(d-p-1)/d}(\log n)^p.$$

To prove (4.24), we will use another partition of $[0,1]^d$. We begin by considering the subcube S of edge length $1 - 2n^{-1/d}$ and centered within $[0,1]^d$. Notice that with high probability there are at most $C n^{(d-1)/d}$ points in the moat $[0,1]^d - S$ and these points contribute at most $C n^{(d-p-1)/d}$ to (4.24).

To complete the proof of (4.24) we have to show

$$\sum_{M_i \in S} |M_i - B_i|^p \le C \cdot n^{(d-p-1)/d} \cdot (\log n)^p.$$

Let F denote a face of $[0,1]^d$. Observe that S is the disjoint union of $Cn^{(d-1)/d}$ rectangular solids which are perpendicular to F, have height $1 - 2n^{-1/d}$, and have a base with diameter $n^{-1/d}$. See Figure 4.2. Geometric considerations show that every such solid contains at most one edge of the graph G_B which is rooted to F. Were there two or more such edges this would contradict optimality, as it would be more efficient to join the points rooted to F with a single edge. Thus the number of points in S which are joined to the boundary of $[0,1]^d$ is at most $Cn^{(d-1)/d}$.

Moreover, the point M_i must be the sample point in the solid which is closest to the boundary. By considering rectangular solids which are perpendicular to the remaining faces of $[0,1]^d$ we may easily conclude that given $M_i \in S$, there is a rectangular solid R such that among all sample points in R, M_i is the one closest to the boundary. Thus for all $M_i \in S$ we have the high probability estimate

$$|M_i - B_i|^p \le C(\log n)^p n^{-p/d}.$$

Since there are as many points in S as there are solids, (4.24) follows.

We now add three types of edges to the trees $T_1, ..., T_N$ which comprise the graph G_B. For all $1 \le i \le N$ insert the edge F_i joining M_i to the nearest point in the grid $G := (G_i)_{i=1}^{n^{(d-1)/d}}$ of regularly spaced points on $\partial[0,1]^d$. See Figure 4.2.

Since each B_i is within $n^{-1/d}$ of a point in G, (4.24) and the triangle inequality imply the high probability estimate

$$(4.25) \qquad S_F^p := \sum_{i=1}^{N} |F_i|^p \le Cn^{(d-p-1)/d}(\log n)^p.$$

Next, for all $1 \le i \le n^{(d-1)/d}$, consider the edge E_i joining G_i to the nearest sample point, say $U_{\sigma(i)}$, where σ is some function with domain $1, 2, ..., n^{(d-1)/d}$ and range contained in $1, 2, ..., n$. There are $n^{(d-1)/d}$ such edges and since each edge length satisfies the high probability bound $|E_i| \le C(\log n/n)^{1/d}$, it follows that

$$(4.26) \qquad S_E^p := \sum_{i=1}^{N} |E_i|^p \le Cn^{(d-p-1)/d}(\log n)^p$$

with high probability.

Figure 4.2. Estimating the lengths of edges joined to the boundary

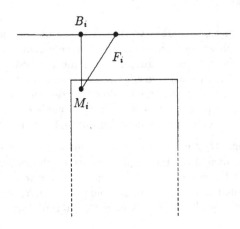

By inserting the two types of edges F_i, $1 \leq i \leq N$, and E_i, $1 \leq i \leq N$, we generate a boundary rooted tree on the the union $G \cup \{U_1, ..., U_n\}$; this tree has disjoint components, say $T_1, ..., T_L$, where $L \leq N$. Given the grid G note that each grid point in G is centered in a $d - 1$ dimensional cube on $\partial[0, 1]^d$. Say that two components are *neighbors* if they contain grid points which are centered in adjacent grid cubes.

The triangle inequality implies that we may tie together any two neighboring components with a third type of edge H, with a length which may be bounded in terms of lengths of edges of the first two types. Let H_i, $i \geq 1$, denote an enumeration of all such edges. Then the H_i, $i \geq 1$, together with the edges in the graph G_B, form a global tree T' through the union $G \cup (U_1, ..., U_n)$. The triangle inequality, (4.25), and (4.26) imply that

$$(4.27) \qquad S_H^p := \sum_{i=1}^{L} |H_i|^p \leq C(S_E^p + S_F^p + n^{(d-p-1)/d}) \leq C n^{(d-p-1)/d} (\log n)^p,$$

where the last inequality holds with high probability. Moreover, by deleting all edges in T' which involve grid points, we form a feasible tree T through the pruned set $(U_1, ..., U_n)$, which shows

$$M^p(n) \leq \sum_{e \in T} |e|^p \leq M_B^p(n) + S_H^p.$$

Now (4.23) follows from the high probability bound (4.27). This completes the proof of Theorem 4.6. $\square$

4.4. Ergodic Theorems, Superadditive Functionals

Euclidean functionals by definition are invariant under translations and it is thus natural to look for an intrinsic ergodic structure yielding laws of large numbers. We pursue this line of inquiry in this section. We will see that a scaled a.s. version of (4.3) is essentially a consequence of a multiparameter subadditive ergodic theorem, one which generalizes Kingman's subadditive ergodic theorem (Kingman (1968,1973)). Such theorems hold for real-valued multiparameter functionals $L := (L(R), \ R \in \mathcal{R}(d))$ defined on a probability space $(\Omega, \mathcal{A}, P)$. We will use these theorems to help identify the limiting constants of Theorems 4.1 and 4.3. Before stating our results, we need a few definitions.

Say that the functional L is *stationary* if for all $m \geq 1$, $R_1, ..., R_m \in \mathcal{R}(d)$ and $u \in (\mathbb{R}^+)^d$, the joint distribution of $L(R_1), ..., L(R_m)$ and $L(R_1 + u), ..., L(R_m + u)$ are the same. Say that L is *bounded* if

$$\sup_n EL([0,n]^d)/n^d < \infty.$$

Say that L is *strongly superadditive* if

$$L(R) \geq \sum_{i=1}^{m} L(R_i),$$

where the rectangles R_i, $1 \leq i \leq m$, form a partition of the rectangle $R \in \mathcal{R}(d)$. Notice that strong superadditivity is stronger than the usual superadditivity (3.3). Strong superadditivity is critical in our search for ergodicity. All superadditive functionals encountered in this monograph are strongly superadditive.

Observe that L is not required to satisfy homogeneity (3.2). There are many examples of stationary strongly superadditive functionals beyond those defined by combinatorial optimization problems. One such example involves the number of clusters in the percolation model. More precisely, if $L(R)$, $R \in \mathcal{R}(d)$, denotes the number of clusters in the percolation model in the rectangle R, then the cluster functional $-L(R)$, $R \in \mathcal{R}(d)$, is strongly superadditive, as observed by Grimmett (1976). Note that $L(R)$, $R \in \mathcal{R}(d)$, is not homogeneous.

Hammersley (1974) provided another example of a strongly superadditive functional. His example, motivated by the statistical theory of liquid-vapor equilibrium, is as follows. Distribute points in $\mathbb{R}^d$ according to a stationary Poisson point process. Fix $r > 0$ and consider a sphere of radius r around each point. For $R \in \mathcal{R}(d)$ let $L(R)$ denote the volume of the spheres whose interiors are wholly in R (we count each element of volume once only, regardless of the number of spheres covering it). Then it is easily checked that $L(R)$ is a bounded, stationary strongly superadditive functional. It is not homogeneous and thus not a Euclidean functional.

A third example of a strongly superadditive functional consists of a modification of an example of Hammersley and Welsh (1965), which we learned from Smythe (1976). Let straight lines be distributed on the plane uniformly and independently at random (in other words their directions are uniformly and independently distributed between 0 and 2π, and their perpendicular distances from the origin are the points of a Poisson process on the positive reals). Let $L(R)$, $R \in \mathcal{R}(2)$, be the number of polygons of some given class (acute triangles, for example) which intersect R. Then $-L(R)$, $R \in \mathcal{R}(2)$, is strongly superadditive but not homogeneous.

To understand the behavior of stationary, strongly superadditive functionals on large cubes, we will appeal to the following theorem, stated without proof. This theorem is due to Akcoglu and Krengel (1981) although the L^1 part follows as in Smythe (1976, Theorem 1.1). This result, which is essentially a strong law of large numbers, generalizes Kingman's (1968) deep subadditive ergodic theorem. This generalized ergodic theorem has been used in statistical physics to analyze the behavior of long range spin systems; see e.g. Van Enter and Van Hemmen (1983). In this section $\mathcal{R}(d)$ denotes rectangles in $\mathbb{N}^d$.

Theorem 4.9. *(Akcoglu and Krengel) Let $L := (L(R) : R \in \mathcal{R}(d))$ be a stationary, bounded, strongly superadditive functional defined on $(\Omega, \mathcal{A}, P)$. Then*

$$\lim_{n \to \infty} L([0,n]^d)/n^d = f(L, d)$$

a.s. and in L^1, where $f(L, d) \in L^1(\Omega, \mathcal{A}, P)$. Moreover,

$$Ef(L, d) = \alpha(L, d) = \sup_R \frac{EL(R)}{volume R}.$$

Clearly $\alpha(L, d)$ is positive whenever L is not identically zero. By the assumed boundedness of L, $\alpha(L, d)$ is finite.

We are now positioned to give a second proof of (4.5) for functionals L which are strongly superadditive. We will use Theorem 4.9 heavily.

Proof of (4.5). Let $\Pi := \Pi(1)$ denote a Poisson point process on $(\mathbb{R}^+)^d$ with intensity 1 and put

$$L_B^p(R) := L_B^p(\Pi \cap R, R), \quad R \in \mathcal{R}(d).$$

We need to verify that the functional $L_B^p(R)$, $R \in \mathcal{R}(d)$, satisfies the conditions of Theorem 4.9. Stationarity follows from translation invariance (3.1) and strong superadditivity follows by hypothesis. Since $\Pi \cap [0,n]^d \overset{d}{=} n(U_1, ..., U_N)$, where N is an independent Poisson random variable with parameter n^d, we see that boundedness results from homogeneity (3.2) and Jensen's inequality:

$$\begin{aligned}
EL_B^p([0,n]^d) &= EL_B^p(n(U_1, ..., U_N), [0,n]^d) \\
&= n^p EL_B^p(U_1, ..., U_N, [0,1]^d) \\
&= Cn^p EN^{(d-p)/d} \\
&\leq Cn^d.
\end{aligned}$$

Thus Theorem 4.9 gives the existence of a function $f(L_B^p, d) \in L^1(\Omega, \mathcal{A}, P)$ such that

(4.28)
$$\lim_{n \to \infty} L_B^p([0, n]^d)/n^d = f(L_B^p, d)$$

a.s. and in L^1.

Since
$$\lim_{n \to \infty} EL_B^p([0, n]^d)/n^d = Ef(L_B^p, d) = \alpha(L_B^p, d)$$

we obtain by homogeneity (3.2) a Poissonized version of (4.5):
$$\lim_{n \to \infty} EL_B^p(U_1, ..., U_N)/n^{d-p} = \alpha(L_B^p, d).$$

By smoothness (3.8) the convergence is unaffected if N is replaced by n^d. Easy interpolation arguments and smoothness for L_B^p show that the argument n^d may be replaced by n, which yields precisely (4.5). Moreover, when p lies in the range $1 \leq p < d$, (4.5) is equivalent to the c.c. version but we have to wait until Chapter 6 (Corollary 6.4) to give the details.

We may use Theorem 4.9 to identify the constant $\alpha(L_B^p, d)$ in Theorems 4.1 and 4.3 and to show that it is necessarily positive. Let Π denote a Poisson point process on $\mathbb{R}^d$ with intensity 1 and put

$$L(R) := L_B^p(\Pi \cap R, R),$$

where L_B^p is a smooth superadditive Euclidean functional of order p on $\mathbb{R}^d$. Then Theorems 4.1, 4.3, and 4.9 give the existence of the limit

$$\lim_{n \to \infty} EL(n \cdot [0, 1]^d)/n^d = \lim_{n \to \infty} EL_B^p\left(\Pi \cap (n \cdot [0, 1]^d), n \cdot [0, 1]^d\right)/n^d$$
$$= \lim_{n \to \infty} EL_B^p(U_1, ..., U_{N(n)}, [0, 1]^d)/n^{(d-p)/d}$$

and show that it equals the *spatial constant*

(4.29)
$$\alpha(L_B^p, d) = \sup_{R \in \mathcal{R}(d)} \frac{EL_B^p(\Pi \cap R, R)}{\text{volume} R}.$$

By assumption $L_B^p(F, R) > 0$ when $\text{card} F > 0$ and thus (4.29) shows that the spatial constant $\alpha(L_B^p, d)$ is positive. Spatial constants are the multiparameter analogs of the time constants appearing in one-dimensional subadditive theory.

Remarks On the Limiting Constants:

(i) In the statement of Theorem 4.9, the number N of sample points in $[0, n]^d$ and the volume V of $[0, n]^d$ tend to infinity but in such a way that the particle number density N/V essentially remains finite. Borrowing a term from statistical mechanics, we say that $f(L_B^p, d)$ is the infinite volume limit or thermodynamic limit for the functional L_B^p.

(ii) Since the constant $\alpha(L, d)$ given by Theorem 4.9 is called the spatial constant for the functional L we interpret the constant $\alpha(L_B^p, d)$ of Theorem 4.1 as the spatial constant for the Euclidean functional L_B^p. We now review some of the estimates for the constants $\alpha(S^p, d)$, $\alpha(M^p, d)$, and $\alpha(T^p, d)$. Little is known concerning the exact values of these mysterious constants but perhaps changing the Euclidean norm to another norm may offer a path towards progress.

$\alpha(S^1, 2)$ is shown in Papadimitriou (1978a) to lie in the interval $[0.25, 0.40106]$ and based on Monte Carlo experiments it is conjectured that $\alpha(S^1, 2) = 0.35$. Bertsimas and van Ryzin (1990) use Crofton's method to show that

$$\frac{1}{2\pi^{1/2}}\Gamma(1/d + 1)\Gamma(d/2 + 1)^{1/d} \leq \alpha(S^1, d)$$

$$\leq \frac{d}{(2d - 1)\pi^{1/2}}\Gamma(1/d + 1)\Gamma(d/2 + 1)^{1/d}2^{1/d}$$

from which it follows that as $d \to \infty$

$$\alpha(S^1, d) \sim \frac{1}{2}(d/2\pi e)^{1/2}.$$

Concerning the MST, Gilbert (1965) shows that $\alpha(M^1, 2)$ is bounded by $2^{-1/2} \approx 0.707$ and he obtains $\alpha(M^1, 2) \approx 0.68$ based on experimental evidence. Bertsimas and van Ryzin (1990) show that

$$\frac{1}{\pi^{1/2}}\Gamma(1/d + 1)\Gamma(d/2 + 1)^{1/d} \leq \alpha(M^1, d) \leq \frac{1}{\pi^{1/2}}\Gamma(1/d + 1)\Gamma(d/2 + 1)^{1/d}2^{1/d}$$

from which it follows that as $d \to \infty$

$$\alpha(M^1, d) \sim (d/2\pi e)^{1/2}.$$

More generally they show for all $p < d$ that

$$\alpha(M^p, d) \sim (d/2\pi e)^{p/2}.$$

The constant $\alpha(M^p, d)$ is also treated by Avram and Bertsimas (1992), who identify $\alpha(M^p, d)$ with an infinite series. Using their infinite series identification, they show that $\alpha(M^1, 2)$ is bounded below by 0.600822 which improves upon the lower bound of 0.5 given by Bertsimas and van Ryzin (1990).

$\alpha(T^1, 2)$ is estimated by Marks (1948) and is known to be bounded by $0.62 < \alpha(T^1, 2) < 0.93$ (Beardwood, Halton, Hammersley (1959, p. 302, and Lemma 3.10)); it is found numerically as $\alpha(T^1, 2) = 0.749$ (Bonomi and Lutton (1984)). Rhee (1992) applies an estimate of Talagrand (1992) to show that

$$|\alpha(T^1, d) - (d/2\pi e)^{1/2}| \le K(\log d/d)$$

which shows that as $d \to \infty$ we have $\alpha(T^1, d) \sim \alpha(M^1, d)$.

(iii) The identification $Ef(L, d) = \alpha(L, d)$ may be seen from Smythe's (1976) mean ergodic theorem, which proves a mean version of (4.28) using a slightly different form of superadditivity.

4.5. Concluding Remarks

This chapter describes a general framework for proving limit theorems for the classic problems of combinatorial optimization. We have focussed on the prototypical problems involving the TSP, MST, and minimal matching functionals. Our approach is useful for other problems, including the Steiner MST problem, the semi-matching problem (Chapter 8), the minimal triangulation problem (Chapter 9), and the k-median problem of Steinhaus (Chapter 10). The approach proves useful for finding asymptotics of functionals L^p with pth power-weighted edges, where p ranges over both the critical range $p \ge d$ as well as the usual range $1 \le p < d$. The method described here consists of two main steps:

(i) use the superadditive structure of the canonical boundary functional L_B^p to deduce the asymptotics of L_B^p and

(ii) show that the boundary functional L_B^p is close to L^p to deduce identical asymptotics for L^p.

Notes and References

1. The difference between the incrementing model and independent model of problem generation was not explicitly recognized until Weide's thesis (1978). The classic a.s. limit results of Beardwood, Halton, and Hammersley (1959) and Steele (1981a) hold for the incrementing model. Steele (1981b) obtained the first limit result for the independent TSP model.

2. This chapter considered limit theorems for Euclidean functionals $L^p(X_1, ..., X_n)$ where X_i, $i \geq 1$, are i.i.d. with the uniform distribution on the unit cube $[0,1]^d$. In the sequel we will consider limit theorems for $L^p(X_1, ..., X_n)$, where the X_i, $i \geq 1$, have a general distribution. When the X_i, $i \geq 1$, have a "self similar" distribution, Lalley (1990) showed that the asymptotics of L^p may involve growth rates which are different from the usual rate of $n^{(d-p)/d}$ and which may involve periodic functions.

3. Returning to the context of Theorem 4.6, recall that $M^p(n) := M^p(U_1, ..., U_n)$. Kesten and Lee (1996) showed an asymptotic result which is a little more precise than (4.14). They showed

$$\lim_{n \to \infty} \left((n+1)^{p/d} E M^p(n+1) - n^{p/d} E M^p(n) \right) = C(d, p),$$

where $C(d, p)$ is a constant depending only on p and d. They also showed that $(M^p(n) - E M^p(n)) / n^{(d-2p)/2d}$ has a limiting normal distribution. See Lee (1997a,b) for more general results, especially central limit theorems over non-uniform samples. A deep and challenging open problem involves showing analogous limiting behavior for the TSP and minimal matching functionals.

4. The Akcoglu and Krengel Theorem 4.9 strengthens previous multiparameter superadditive ergodic theorems given earlier by Smythe (1976) and Nguyen (1979).

5. An interesting open problem involves using the Aldous and Steele objective method to prove Theorem 4.7 and thereby establish a theoretical value of the limiting constant $\alpha(T_B^d, d)$. McElroy (1997) uses the method of boundary functionals to prove the analog of Theorems 4.6 and 4.7 for the minimal matching and semi-matching functionals.

6. Krengel and Pyke (1987) proved a uniform version of Theorem 4.9 over general averaging sets. Yukich (1996b) used this to deduce uniform a.s. limit results for the TSP functional.

7. Classical multiparameter ergodic theorems (see e.g. Dunford (1951) and Zygmund (1951)) prove laws of large numbers for arrays of random variables $\{X_j\}_{j \in \mathbb{N}^d}$. Instead of assuming boundedness of L, these theorems require a moment condition of the form $E\left(|X_1|(\log|X_1|)^{d-1}\right) < \infty$. Proofs of these theorems (see e.g. Krengel (1985)) may be deduced by simple inductive arguments from the 1-parameter case. The first multiparameter pointwise ergodic theorems go back to Wiener (1939).

5.   RATES OF CONVERGENCE AND HEURISTICS

In this chapter we show that boundary functionals L_B provide the ideal tools for finding rates of convergence of the means $EL(U_1, ..., U_n)$ as well as for analyzing the performance of easily computed partitioning heuristics L_H which approximate L. In both instances, boundary functionals provide a simple approach to problems which have been traditionally attacked by less powerful and less natural bare-hands methods.

5.1.   Rates of Convergence in The Basic Limit Theorem

Chapter 4 provided basic limit theorems for superadditive Euclidean functionals. This chapter examines the rates of convergence of the means of these functionals.

Subadditivity of a functional L is not enough to give rates of convergence. Subadditivity only yields one sided estimates whereas rate results require two sided estimates. However, if the functional L can be made superadditive by appropriately modifying it then we can usually extract rates of convergence. This idea is widely known and discussed in Hammersley (1974), for example. It should be no surprise that boundary functionals L_B provide exactly the right type of modification of L which we are looking for.

The next result shows that if subadditive Euclidean functionals L^p are close in mean (3.16) to the associated superadditive Euclidean functional L_B^p, namely if

$$(5.1) \qquad |EL^p(U_1, ..., U_n) - EL_B^p(U_1, ..., U_n)| \leq C(n^{(d-p-1)/d} \vee 1),$$

then we may find rates of convergence for $EL^p(U_1, ..., U_n)$. Since the TSP, MST, and minimal matching functionals satisfy closeness in mean ($p \neq d-1, d \geq 3$) the following theorem immediately provides rates of convergence for our three prototypical examples.

Theorem 5.1. *(rates of convergence of means). Suppose that L^p and L_B^p are subadditive and superadditive Euclidean functionals of order p, respectively, and that they satisfy the close in mean approximation (5.1). If N is an independent Poisson random variable with parameter n, then for all $d \geq 2$ and $1 \leq p < d$ we have*

$$(5.2) \qquad |EL^p(U_1, ..., U_N) - \alpha(L_B^p, d)n^{(d-p)/d}| \leq C\left(n^{(d-p-1)/d} \vee 1\right).$$

Moreover, if L^p is smooth (3.8), then for $d \geq 2$ and $1 \leq p < d$ we have

$$(5.3) \qquad |EL^p(U_1, ..., U_n) - \alpha(L_B^p, d)n^{(d-p)/d}| \leq C\left(n^{(d-p)/2d} \vee n^{(d-p-1)/d}\right).$$

Proof. For all $n \in \mathbb{N}$ set $\phi(n) := EL^p(U_1, ..., U_{N(n)})$, where $N(n)$ is a Poisson random variable with parameter n and which is independent of the U_i, $i \geq 1$. It follows from translation invariance (3.1), homogeneity (3.2), and subadditivity (3.5) that whenever m is a power of 2 we have

$$\phi(nm^d) \leq m^{-p} \sum_{i=1}^{m^d} \phi(n) + Cm^{d-p}$$
$$= m^{d-p}\phi(n) + Cm^{d-p}.$$

Dividing by $(nm^d)^{(d-p)/d}$ yields the homogenized relation

$$\frac{\phi(nm^d)}{(nm^d)^{(d-p)/d}} \leq \frac{\phi(n)}{n^{(d-p)/d}} + \frac{C}{n^{(d-p)/d}}.$$

As in (4.6) it follows that the limit of the left side as m tends to infinity exists and equals $\alpha(L_B^p, d)$. Thus

$$\frac{\phi(n)}{n^{(d-p)/d}} - \alpha(L_B^p, d) \geq \frac{-C}{n^{(d-p)/d}}$$

or simply

(5.4) $$\phi(n) - \alpha(L_B^p, d)n^{(d-p)/d} \geq -C.$$

Setting $\phi_B(n) := EL_B^p(U_1, ..., U_{N(n)})$ and exploiting the superadditivity of L_B in the same way that we exploited the subadditivity of L, we effortlessly obtain the companion estimate to (5.4) where we may now let $C = 0$:

(5.5) $$\phi_B(n) - \alpha(L_B^p, d)n^{(d-p)/d} \leq 0.$$

By the assumed closeness in mean (5.1), we have by Fubini's theorem and independence

$$|\phi_B(n) - \phi(n)| \leq E_N|E_U L_B^p(U_1, ..., U_N) - E_U L^p(U_1, ..., U_N)|$$
$$\leq E_N(N^{(d-p-1)/d} \vee 1)$$
$$\leq C(n^{(d-p-1)/d} \vee 1),$$

where E_N and E_U denote the expectation with respect to the random variables N and U, respectively. Now (5.1) follows from (5.4) and (5.5). Finally, the de-Poissonized version (5.3) is a simple consequence of smoothness:

$$|EL^p(U_1, ..., U_N) - EL^p(U_1, ..., U_n)| \leq CE(|N - n|^{(d-p)/d})$$
$$\leq Cn^{(d-p)/2d}.$$

This completes the proof of Theorem 5.1. $\square$

5.2. Sharper Rates of Convergence in the Basic Limit Theorem

The proof of Theorem 5.1 is remarkably simple. If it has a shortcoming, it is only that de-Poissonizing introduces an extra error $Cn^{(d-p)/2d}$. We may remove this error whenever L satisfies what Steele calls an "add-one bound" of the type

(5.6) $$|EL^p(U_1, ..., U_{n+1}) - EL^p(U_1, ..., U_n)| \leq Cn^{-p/d}.$$

The improved version of Theorem 5.1 takes the following form:

Theorem 5.2. *(rates of convergence of means). Suppose that L^p and L^p_B are subadditive and superadditive Euclidean functionals of order p, respectively, and that they satisfy the close in mean approximation* (5.1) *and the "add-one bound"* (5.6). *Then for all $d \geq 2$ and $1 \leq p < d$ we have*

$$(5.7) \qquad |EL^p(U_1,...,U_n) - \alpha(L^p_B,d)n^{(d-p)/d}| \leq C\left(n^{(d-p-1)/d} \vee 1\right).$$

Proof. Let N denote an independent Poisson random variable with parameter n and follow the proof of Theorem 5.1 up to the Poissonized estimates (5.4) and (5.5). Conditioning on N we show that the add-one bound (5.6) leads to the de-Poissonized estimate

$$(5.8) \qquad |EL^p(U_1,...,U_n) - EL^p(U_1,...,U_N)| \leq Cn^{1/2-p/d},$$

which will be enough to complete the proof of Theorem 5.2 since $n^{1/2-p/d} \leq n^{(d-p-1)/d}$ holds for all $n \geq 1$, $d \geq 2$, and $1 \leq p < d$.

Now to prove (5.8), we use the decomposition

$$|EL^p(U_1,...,U_n) - EL^p(U_1,...,U_N)|$$
$$\leq |E(L^p(U_1,...,U_n) - L^p(U_1,...,U_N)) \cdot 1_{\{0 \leq N < n/2, \, N > 3n/2\}}| +$$
$$+ |E(L^p(U_1,...,U_n) - L^p(U_1,...,U_N)) \cdot 1_{\{n/2 \leq N \leq 3n/2\}}|$$
$$:= I + II.$$

Using the growth bound $L^p(U_1,...,U_n) \leq C_2 n^{(d-p)/d}$ and the fact that $N - n$ has exponential tails, we obtain the bound $I = O(n^{1/2-p/d})$. By the assumed add-one bound (5.6), term II is bounded by

$$\leq E_N |E_U(L^p(U_1,...,U_n) - L^p(U_1,...,U_N)) \cdot 1_{\{n/2 \leq N \leq 3n/2\}}|$$
$$\leq C\sum_{k=n+1}^{3n/2}(k-n)n^{-p/d}P\{N=k\} + C\sum_{k=n/2}^{n}(n-k)n^{-p/d}P\{N=k\}$$
$$\leq Cn^{-p/d}E|N-n|$$
$$\leq Cn^{1/2-p/d}.$$

Thus (5.6) leads to (5.8) and the proof of Theorem 5.2 is complete. $\square$

Rate results for Euclidean functionals L thus follow once the "add-one bound" (5.6) is satisfied. When L is the MST functional, then (5.6) is satisfied, as shown by Redmond and Yukich (1996). However, if L is the minimal matching or TSP functional, then it is unclear whether (5.6) is satisfied.

On the other hand, in the absence of (5.6), one can still obtain rates of convergence. When $p = 1$ and $d = 2$, for example, Alexander (1994) shows directly that the minimal matching functional satisfies the de-Poissonization estimate

$$|EL^p(U_1,...,U_n) - EL^p(U_1,...,U_N)| = O(1)$$

and in this way obtains the rate result (5.3).

5.3. Optimality of Rates

In general, the rate results (5.1) and (5.3) cannot be improved. This is illustrated in dimension 2 by the following theorems, proved by Jaillet (1993) and Rhee (1994a), respectively. Rhee's theorem settles a conjecture attributed to Karp (1977). Jaillet (1995) presents a fine overview of rate results and we refer to this paper for complete details. Throughout this section, $(U_i)_{i\geq 1}$ denotes an i.i.d. sequence of uniformly distributed random variables with values in the unit square and $N := N(n)$ denotes an independent Poisson random variable with parameter n. It is not clear whether the following inequalities hold when N is replaced by its parameter n.

Theorem 5.3. (optimal rates of convergence for the MST on the square). There is a universal constant C and $n_o \in \mathbb{N}$ such that for all $n \geq n_o$

$$(5.9) \qquad EM(U_1, ..., U_N)/n^{1/2} \geq \alpha(M_B, 2) + Cn^{-1/2}.$$

Theorem 5.4. (optimal rates of convergence for the TSP on the square). There is a universal constant C and $n_o \in \mathbb{N}$ such that for all $n \geq n_o$

$$(5.10) \qquad ET(U_1, ..., U_N)/n^{1/2} \geq \alpha(T_B, 2) + Cn^{-1/2}.$$

The idea behind the estimate (5.9) involves showing that

$$(5.11) \qquad EM(U_1, ..., U_{N(4n)}) \leq 2EM(U_1, ..., U_{N(n)}) - C$$

for some constant $C > 0$. This bound is somewhat surprising since heuristically one might expect that scaling arguments would show the approximate equivalence $EM(U_1, ..., U_{N(4n)}) \sim 2EM(U_1, ..., U_{N(n)})$.

Iterating (5.11) leads to

$$EM(U_1, ..., U_{N(4^m n)}) \leq 2^m EM(U_1, ..., U_{N(n)}) - C(2^m - 1).$$

Dividing by $(4^m n)^{1/2}$ and letting m tend to infinity gives (5.9). A similar approach yields (5.10). The complete proof of (5.9) and (5.10) is too involved to present here and we refer to Jaillet (1993) and Rhee (1994a), respectively, for the details. Steele (1997, Chapter 3) presents a fine exposition of the inequality (5.11).

It is not difficult to prove the analogs of Theorems 5.3 and 5.4 for the canonical boundary functionals M_B and T_B, respectively. In fact, by establishing the lower bound

$$(5.12) \qquad EM_B(U_1, ..., U_{N(4n)}) \geq 2EM_B(U_1, ..., U_{N(n)}) + C$$

we may prove a companion result to Theorem 5.3:

Theorem 5.5. *(optimal rates of convergence for the boundary MST functional on the square). There is a universal constant C and $n_o \in \mathbb{N}$ such that for all $n \geq n_o$*

$$(5.13) \qquad EM_B(U_1, ..., U_N)/n^{1/2} \leq \alpha(M_B, 2) - Cn^{-1/2}.$$

An interesting open question concerns rates of convergence for the MST functional on the torus, namely the unit square equipped with the flat metric. The MST functional M_T on the torus T is defined as follows: for $F \subset T$, $M_T(F)$ denotes the length of the minimal spanning tree through F, where distances are measured in terms of the flat metric. Since M_T is related to the boundary MST M_B and the standard MST M by $M_B \leq M_T \leq M$, M_T clearly admits the rate of convergence

$$(5.14) \qquad |EM_T(U_1, ..., U_n) - \alpha(M_B, 2)n^{1/2}| \leq C.$$

However, it is not clear that this is the exact rate of convergence. In other words, are the rates of convergence expressed by (5.14) optimal?

5.4. Analysis of Partitioning Heuristics

As we saw in Chapter 4, the asymptotics for the TSP

$$\lim_{n \to \infty} T(U_1, ..., U_n)/n^{(d-1)/d} = \alpha(T, d) \quad a.s.$$

lead Karp (1976, 1977) to find efficient methods for approximating the length $T(U_1, ..., U_n)$ of the shortest path through i.i.d. uniformly distributed random variables $U_1, ..., U_n$ on the unit square. Karp developed the "fixed dissection algorithm" which provides a simple heuristic $T_H(U_1, ..., U_n)$ having the property that $T_H(U_1, ..., U_n)/T(U_1, ..., U_n)$ converges completely to 1 and which moreover has polynomial mean execution time.

The fixed dissection algorithm consists of dividing the unit cube $[0, 1]^d$ into m^d congruent subcubes $Q_1, ..., Q_{m^d}$, finding the shortest tour T_i of length $T(\{U_1, ..., U_n\} \cap Q_i)$ on each of the subcubes, constructing a tour T which links representatives from each T_i, and then deleting excess edges to generate a grand (heuristic) tour through $U_1, ..., U_n$ having length $T_H(U_1, ..., U_n)$. See Figure 5.1.

Using Karp's seminal work as a guide, Karp and Steele (1985) show via elementary methods that the partitioning heuristic T_H is ϵ-optimal with probability one:

Theorem 5.6. *(Karp and Steele, 1985) If $m^d := n/\sigma$, where σ is an unbounded increasing function of n, then for all $\epsilon > 0$*

$$\sum_{n=1}^{\infty} P\left\{ \frac{T_H(U_1,...,U_n)}{T(U_1,...,U_n)} \geq 1+\epsilon \right\} < \infty.$$

Theorem 5.6 thus shows that the ratio of the lengths of the heuristic tour and the optimal tour converges completely to 1. Given the computational complexity of the TSP, it is remarkable that the optimal tour length is so well approximated by a sum of individual tour lengths, where the sum has polynomial mean execution time.

Figure 5.1.   Karp's TSP heuristic: the grand tour is constructed from the local tours by inserting the dashed edges and deleting the dotted edges

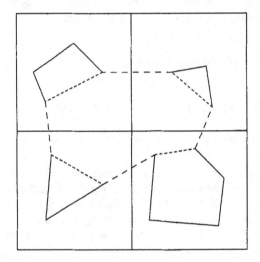

The Karp and Steele result is an example of a general phenomenon: given a Euclidean functional L, there is a heuristic L_H which closely approximates it in the sense of Theorem 5.6. We show that the general framework of subadditive and superadditive Euclidean functionals L and L_B, respectively, permits a strikingly easy analysis of a partitioning heuristic L_H which shares many of the same pleasant features as Karp's partitioning heuristic for the TSP. The use of boundary functionals, especially their closeness property to the standard functional, provides a simplifying tool.

The results of this section essentially show that Euclidean functionals L^p can be approximately represented as a sum of i.i.d. local functionals $L^p(F \cap Q_i, Q_i)$, $1 \leq i \leq m^d$. The upcoming Lemma 5.8 shows that $L^p(F)$ may be approximated by the sum of the independent local functionals $L^p(F \cap Q_i, Q_i)$ on the subcubes Q_i, $1 \leq i \leq m^d$, plus a correction term which is deterministically small compared to $n^{(d-p)/d}$.

The correction term may be regarded as a way of measuring the interactions between the local functionals on the m^d subcubes, a point of view which is valuable in statistical mechanics. The correction term is small enough to yield a strong law of large numbers, but not quite small enough to yield a central limit theorem.

Let us now fix our ideas. Let L^p be a subadditive Euclidean functional associated with the solution to an optimization problem. We assume $L^p \geq L_B^p$. Given $F \subset [0,1]^d$ and $L^p(F)$, consider the feasible solution obtained by solving the optimization problem on the subcubes Q_i, $1 \leq i \leq m^d$, and then adding and deleting edges in the resulting graph to obtain a global solution on the set F. This feasible solution, which we call the canonical heuristic, has a length denoted by $L_H^p(F, m^d)$. $L_H^p(F, m^d)$ is the sum of the local functionals $L^p(F \cap Q_i, Q_i)$, $1 \leq i \leq m^d$, plus a correction term which we implicitly assume is bounded by $C_1 m^{d-p}$ whenever m is a power of 2. Thus we assume that L_H^p satisfies for all $F \subset [0,1]^d$

$$(5.15) \qquad L^p(F) \leq L_H^p(F, m^d) \leq \sum_{i=1}^{m^d} L^p(F \cap Q_i, Q_i) + C_1 m^{d-p}.$$

When L is the TSP functional, for example, then L_H is Karp's (1976,1977) heuristic.

It will be convenient to let the number m^d of subcubes depend on the cardinality of F, which for brevity we denote by $|F|$. This approach is similar to that of Karp (1976,1977). To make this precise, let $\sigma := \sigma(n)$ denote a function of n such that $\sigma(n)$ increases up to infinity and $1 < \frac{n}{\sigma(n)} = 2^{d \cdot j(n)}$ for some non-decreasing sequence of integers $j(n)$, $n \geq 1$. For such functions σ we will consider heuristics $L_\sigma^p(F)$ of the form

$$L_\sigma^p(F) := L_H^p\left(F, \frac{|F|}{\sigma(|F|)}\right).$$

Thus the heuristic $L_\sigma^p(F)$ subdivides the unit cube into $\frac{|F|}{\sigma(|F|)}$ subcubes. If we set $m^d := \frac{|F|}{\sigma(|F|)}$ we obtain from (5.15):

$$(5.16) \qquad L_\sigma^p(F) \leq \sum_{i=1}^{|F|/\sigma(|F|)} L^p(F \cap Q_i, Q_i) + C_1\left(\frac{|F|}{\sigma(|F|)}\right)^{(d-p)/d}.$$

In order to proceed we will need to make a closeness assumption on the Euclidean functional L. Suppose that L^p is a Euclidean functional which is pointwise close to its boundary functional L_B^p in the sense that for all $F \subset [0,1]^d$ we have

$$(5.17) \qquad |L^p(F) - L_B^p(F)| \leq C\left((\mathrm{card}F)^{(d-p-1)/(d-1)} \vee \log(\mathrm{card}F)\right).$$

By Lemma 3.7 we know that the TSP, MST, and minimal matching functionals satisfy (5.17).

We now state a deterministic result which shows that the heuristic solutions $L_\sigma^p(F)$ have a length which is close to that of the optimal solution.

Lemma 5.7. *Assume that L^p and L^p_B are subadditive and superadditive Euclidean functionals of order p, respectively, and that they satisfy pointwise closeness (5.17). Then the heuristic L^p_σ satisfies*

(5.18)
$$\sup_{|F|=n} |L^p(F) - L^p_\sigma(F)| = o(n^{(d-p)/d}).$$

Proof. Consider $F \subset [0,1]^d$, where $\text{card}F = n$. For ease of notation set $m = (n/\sigma(n))^{1/d}$ and note that m is a power of 2 by hypothesis, i.e., $m = 2^{j(n)}$. For ease of presentation we take $p < d-1$; therefore the log term in (5.17) may be ignored and we obtain

(5.19)
$$|L^p(F, [0,1]^d) - L^p_B(F, [0,1]^d)| \leq Cn^{(d-p-1)/(d-1)}.$$

Since m is a power of 2 we have by (5.16) and (5.19)

$$L^p(F) \leq L^p_\sigma(F)$$

$$\leq \sum_{i=1}^{m^d} L^p(F \cap Q_i, Q_i) + C_1 m^{d-p}$$

$$\leq \sum_{i=1}^{m^d} \left(L^p_B(F \cap Q_i, Q_i) + m^{-p} C(\text{card}(F \cap Q_i))^{(d-p-1)/(d-1)} \right) + C_1 m^{d-p}$$

$$\leq L^p(F) + m^{-p} \sum_{i=1}^{m^d} C \left(\text{card}(F \cap Q_i) \right)^{(d-p-1)/(d-1)} + C_1 m^{d-p}.$$

Thus we have

(5.20) $$|L^p(F) - L^p_\sigma(F)| \leq m^{-p} \sum_{i=1}^{m^d} C \left(\text{card}(F \cap Q_i) \right)^{(d-p-1)/(d-1)} + C_1 m^{d-p}.$$

The above sum is largest when $\text{card}(F \cap Q_i) = n/m^d$, $1 \leq i \leq m^d$. It is thus bounded by $Cm^{p/(d-1)}n^{(d-p-1)/(d-1)}$. Using the definition of m we arrive at the estimate

(5.21) $$|L^p(F) - L^p_\sigma(F)| \leq C\sigma^{-p/d(d-1)}n^{(d-p)/d} + C_1\sigma^{(p-d)/d}n^{(d-p)/d}.$$

Thus (5.21) tells us that the heuristic $L^p_\sigma(F)$ is larger than $L^p(F)$ by a quantity which is deterministically small compared to $n^{(d-p)/d}$. This completes the proof of Lemma 5.7. □

By Lemma 5.7 the asymptotic behavior of the scaled heuristic

(5.22)
$$\frac{L^p_\sigma(X_1, ..., X_n)}{n^{(d-p)/d}}$$

coincides with the asymptotic behavior of the scaled Euclidean functional

(5.23)
$$\frac{L^p(X_1, ..., X_n)}{n^{(d-p)/d}},$$

where X_i, $i \geq 1$, are i.i.d. random variables with values in $[0,1]^d$. While we have not treated the asymptotics of Euclidean functionals over non-uniform samples we will show in Chapters 6 and 7 that the functional (5.23) converges completely to a positive constant whenever the law of X_1 has a continuous part. It will thus follow that (5.22) also converges completely to a constant. Admitting the complete convergence of (5.22) we can now extend Karp and Steele's result to general Euclidean functionals over general sequences of random variables:

Theorem 5.8. (the heuristic L_σ^p is ϵ-optimal) Let L^p and L_B^p be subadditive and superadditive Euclidean functionals of order p, respectively. Assume that they are pointwise close (5.17). Then for all $\epsilon > 0$ and all i.i.d. sequences X_i, $i \geq 1$, of random variables with a continuous part, the heuristic L_σ^p is ϵ-optimal:

(5.24)
$$\sum_{n=1}^{\infty} P \left\{ \frac{L_\sigma^p(X_1, ..., X_n)}{L^p(X_1, ..., X_n)} \geq 1 + \epsilon \right\} < \infty.$$

Proof. We will assume that (5.23) converges completely to a positive constant C, a fact which will be proved in Chapters 6 and 7. By Lemma 5.7 it follows that the scaled heuristic (5.22) converges completely to C as well. It therefore follows by standard arguments that the ratio of (5.22) to (5.23) converges completely to 1. This is precisely (5.24). $\square$

We now show that the expected execution time for the partitioning heuristic L_σ^p is polynomially bounded under some weak assumptions. This feature, together with (5.19), show that L_σ^p has all the properties of Karp's heuristic.

The time required to compute the heuristic $L_\sigma^p(U_1, ..., U_n)$ is bounded by

$$T_n := \sum_{i=1}^{\frac{n}{\sigma(n)}} f(N_i),$$

where $N_i := \operatorname{card}\{Q_i \cap \{U_1, ..., U_n\}\}$, $1 \leq i \leq n/\sigma(n)$, and where $f(N)$ denotes a bound on the time needed to compute $L^p(F)$, $\operatorname{card} F = N$. If we assume that f exhibits polynomial growth, say $f(x) = Ax^B 2^x$ for some constants A and B, as would be the case for the TSP, then since the N_i, $1 \leq i \leq n/\sigma(n)$, are binomial random variables, straightforward calculations show that

$$ET_n \leq 4An(\sigma(n))^{B-1} \exp(\sigma(n))$$

see e.g. Karp and Steele (1985).

We now want to choose $\sigma(n)$ in such a way that we minimize the expected computation time T_n. We notice that there is a non-decreasing sequence $j = j(n)$, $n \geq 1$, such that

$$C \log n \leq \frac{n}{2^{dj}} \leq \log n$$

for some constant $C < 1$. Here and henceforth we set $\sigma(n) = \frac{n}{2^{dj}}$ so that $\sigma(n) \leq \log n$. This value of $\sigma(n)$ implies that the expected execution time for the heuristic L_σ^p is $O(n^2 \log^{B-1} n)$.

Using this value of $\sigma(n)$ in (5.21), the above discussion may be summarized by the following general result, which applies to the TSP and other optimization functionals. As before we write $|F|$ to denote card$F$.

Theorem 5.9. *With $\sigma := \sigma(n)$ as above, let L_σ^p denote the heuristic associated with the subadditive Euclidean functional L^p. If L^p is pointwise close (5.17) to the boundary functional L_B^p, then the heuristic L_σ^p closely approximates L^p*

$$|L^p(F) - L_\sigma^p(F)| \leq C \left((\log |F|)^{-p/d(d-1)} + (\log |F|)^{(p-d)/d} \right) |F|^{(d-p)/d}$$

and is ϵ-optimal (5.24) with probability one. Moreover, if $L^p(F)$ may be computed in time bounded by $A(|F|)^B \cdot 2^{|F|}$, then the expected execution time for $L_\sigma^p(U_1, ..., U_n)$ is $O\left(n^2 (\log n)^{B-1}\right)$.

Clearly, by making a different choice of σ, e.g. $\sigma = \sigma(n) = \log \log n$, we can reduce the expected execution time at the expense of increasing the bound for $|L^p(F) - L_\sigma^p(F)|$.

5.5. Concluding Remarks, Open Questions

In this chapter we have already indicated a few open problems. We close with a list of several more problems and remarks.

1. When $p = 1$ and L is the TSP functional, the rate (5.3) was conjectured by Beardwood, Halton, and Hammersley (1959). Alexander (1994) obtained a result similar to (5.3) without the simplifying use of boundary functionals. He instead requires that L satisfy several hypotheses beyond the minimal ones (3.1)-(3.3). These hypotheses are shown to give a form of superadditivity similar to (3.3), but with a non-negligible correction term. Alexander applies his results to the TSP functional and in this way was the first to settle the rate conjecture of Beardwood, Halton, and Hammersley.

2. We have already seen from (5.11) that the estimate

$$|EM(U_1, ..., U_{N(4n)}) - 2EM(U_1, ..., U_{N(n)})| = O(1)$$

cannot be improved. Thus one wouldn't expect that the estimate

$$|M(U_1, ..., U_{N(4n)}) - \sum_{i=1}^{4} M(U_1, ..., U_{N(4n)} \cap Q_i)| = O_P(1)$$

could be improved, where Q_1, Q_2, Q_3, Q_4 represents the usual partition of $[0,1]^2$ into four congruent subsquares. Yet it is conceivable that there is a constant ψ such that

$$|M(U_1, ..., U_{N(4n)}) - \sum_{i=1}^{4} M(U_1, ..., U_{N(4n)} \cap Q_i) - \psi| = o_P(1).$$

Clearly a similar estimate might hold for the TSP functional.

The following intriguing conjecture, due to Rhee (1994a), formalizes these remarks. Her conjecture is based on an analysis of the behavior of the TSP functional near the boundary of the square. Let Π be a Poisson point process of constant intensity λ on the unit square.

Conjecture 5.10. *(Rhee, 1994a) There is a constant ψ such that given $\epsilon > 0$ for large enough λ we have*

(5.25) $$P\{|T(\Pi) - \sum_{i=1}^{4} T(\Pi \cap Q_i, Q_i) - \psi| \geq \epsilon\} \leq \epsilon.$$

Were it true, this conjecture would show that the centered and Poissonized TSP functional $T(\Pi) - ET(\Pi)$ is asymptotically normal, i.e., satisfies a central limit theorem. To see this, one only needs to iterate (5.25) k times to approximately represent $T(\Pi) - ET(\Pi)$ as a sum of 2^k i.i.d. random variables.

3. Karp (1977) anticipates the general rate results of this chapter and observes that in dimension 2 the TSP functional satisfies

$$ET(U_1, ..., U_n) - \alpha(T_B, 2)n^{1/2} \leq C.$$

This, combined with the simple estimate (5.4), leads to the correct convergence rate for the TSP in dimension 2.

4. Section 5.4 extends and strengthens the results in Yukich (1995b, section 4), which is motivated in part by discussions with D. Bertsimas.

5. It should be possible to find rates of convergence for the power weighted functional L^p, when $p = d$. This would involve using the methods of Chapter 4 and improved subadditive bounds for L^p.

6. Is it possible to refine the rate results (5.2) and (5.3) by finding a function of n, say $\gamma(n)$, such that

$$|EL(U_1, ..., U_N) - \alpha(L_B^1, d)n^{(d-1)/d} - \gamma(n)| = o(1)?$$

6. ISOPERIMETRY AND

CONCENTRATION INEQUALITIES

6.1. Azuma's Inequality

The last chapter provided a rate of convergence for the means of subadditive and superadditive Euclidean functionals L^p and L^p_B, but it didn't tell us how these functionals are concentrated around their means in terms of deviation inequalities. This chapter will show that smoothness conditions lead to sharp concentration estimates via isoperimetric methods. We begin by reviewing some of the deviation estimates which are basic to our subject.

When $d = 2$, Steele (1981b) showed by means of the jackknife inequality of Efron and Stein that the variance of $T(U_1, ..., U_n)$ is bounded independently of n. In fact, writing $T(n)$ for $T(U_1, ..., U_n)$ where U_i, $i \geq 1$, are i.i.d. with the uniform distribution on $[0, 1]^2$, he proved that

$$(6.1) \qquad \sum_{n=1}^{\infty} P\{|T(n)/n^{1/2} - \alpha(T^1_B, 2)| > \epsilon\} < \infty$$

for all $\epsilon > 0$. We recall that (6.1) expresses the *complete convergence* of $T(n)/n^{1/2}$ and is of course stronger than a.s. convergence. Steele (1981b) was motivated to show (6.1) in order to rigorously justify Karp's (1976,1977) algorithm for the traveling salesman problem under the *independent model of problem generation*, a topic described at the beginning of Chapter 4.

The estimate (6.1) provides a weak deviation estimate for the TSP functional about its mean. In this chapter we show that many Euclidean functionals enjoy similar and even more refined concentration inequalities. It might be anticipated that estimates of the type (6.1) hold for Euclidean functionals which are reasonably well controlled and which do not change too much when the underlying vertex set is perturbed. This is indeed the case and we will see shortly that (6.1) admits a generalization to all *smooth* Euclidean functionals.

Deviation estimates which are more refined than (6.1) may often be established via martingale methods, as first discovered by Rhee and Talagrand (1987). This approach has the advantage of versatility and applies to a wide range of Euclidean functionals.

We first recall the martingale difference sequence representation of an arbitrary random variable $X \in L^1(\Omega, \mathcal{A}, P)$. Given a filtration

$$(\emptyset, \Omega) = \mathcal{A}_0 \subset \mathcal{A}_1 \subset ... \subset \mathcal{A}_n = \mathcal{A}$$

of σ-algebras of $\mathcal{A}$, let $E(X|\mathcal{A}_i)$ denote the conditional expectation of X with respect to $\mathcal{A}_i$. For each $1 \leq i \leq n$ define the martingale difference

$$d_i := E(X|\mathcal{A}_i) - E(X|\mathcal{A}_{i-1})$$

so that $X - EX = \sum_{i=1}^{n} d_i$. This generic representation of $X - EX$ will allow us to express the high concentration of X around its mean in terms of the size of the differences d_i.

Let $\|d_i\|_\infty$ denote the essential supremum of d_i. The following fundamental inequality, due to Azuma (1967), provides deviation bounds for the above martingale decomposition.

Theorem 6.1. *(Azuma's inequality) For all $t > 0$*

$$P\{|\sum_{i=1}^{n} d_i| \geq t\} \leq 2\exp\left(\frac{-t^2}{2\sum_{i=1}^{n} \|d_i\|_\infty^2}\right).$$

Proof. We first note that when X is a mean zero random variable such that $|X| \leq 1$ almost surely, then for any real a

$$E\exp(aX) \leq \exp(a^2/2).$$

To see this, note from the convexity of $f(x) = \exp(ax)$ and from $ax = a(1+x)/2 - a(1-x)/2$ that, for any real $|x| \leq 1$ we have

$$\exp(ax) \leq \cosh a + x\sinh a.$$

Taking expectations and using the elementary bound $\cosh x \leq \exp(x^2/2)$ we obtain the claim. It follows for any $i = 1, ..., n$ that

$$E(\exp(ad_i)|\mathcal{A}_{i-1}) \leq \exp(\frac{a^2}{2}\|d_i\|_\infty^2).$$

Iterating this inequality and using the properties of conditional expectation we get

$$E\exp(t\sum_{i=1}^{n} d_i) = E\left(E\exp(t\sum_{i=1}^{n} d_i)|\mathcal{A}_{n-1}\right)$$

$$= E\left(\exp(t\sum_{i=1}^{n-1} d_i)\, E(\exp(td_n)|\mathcal{A}_{n-1})\right)$$

$$\leq E\exp(t\sum_{i=1}^{n-1} d_i)\, \exp(\frac{t^2}{2}\|d_n\|_\infty^2)$$

$$\cdots$$

$$\leq \exp(\frac{t^2}{2}\sum_{i=1}^{n} \|d_i\|_\infty^2).$$

From Markov's inequality we obtain for all $t > 0$

$$P\{\sum_{i=1}^{n} d_i > \lambda\} \leq \exp(-\lambda t)\exp(\frac{t^2}{2}\sum_{i=1}^{n} \|d_i\|_\infty^2).$$

Letting $t = \lambda(\sum_{i=1}^n \|d_i\|_\infty^2)^{-1}$ we obtain

$$P\{\sum_{i=1}^n d_i > \lambda\} \leq \exp\left(\frac{-\lambda^2}{2\sum_{i=1}^n \|d_i\|_\infty^2}\right).$$

Applying this inequality to the sum $-\sum_{i=1}^n d_i$ yields Azuma's inequality as desired. $\square$

There are several ways to refine Azuma's inequality and they depend largely on bounds for $E(d_i^2|\mathcal{A}_{i-1})$ and $\max_{i \leq n} i \|d_i\|_\infty$. We refer to Ledoux and Talagrand (1991) for a complete treatment. Azuma's inequality and the generic representation $X - EX = \sum_{i=1}^n d_i$ are used in many contexts, especially in studying the deviations of $\|\sum_{i=1}^n X_i\|$ where X_i, $i \geq 1$, are independent Banach space valued random variables; see Ledoux and Talagrand (1991).

Our immediate interest is the application of Azuma's inequality to problems in geometric probability. To illustrate, let U_j, $j \geq 1$, be i.i.d. random variables with the uniform distribution on $[0,1]^d$ and let $\mathcal{A}_i$ denote the σ-field generated by the random variables $U_1, ..., U_i$. Given a Euclidean functional L^p, we abbreviate notation and write $L^p(n)$ for $L^p(U_1, ..., U_n)$. Consider the martingale differences

$$d_i := E\left(L^p(n)|\mathcal{A}_i\right) - E\left(L^p(n)|\mathcal{A}_{i-1}\right)$$

and notice that $L^p(n) - EL^p(n)$ admits the martingale decomposition

$$L^p(n) - EL^p(n) = \sum_{i=1}^n d_i.$$

When L is the shortest tour functional T, then the martingale increments satisfy the bound

$$\|d_i\|_\infty \leq C(d)(n - i + 1)^{-1/d}, \quad d \geq 2,$$

as shown by Rhee and Talagrand (1987). From this they easily obtain a deviation inequality for $T(n)$:

$$(6.2) \qquad P\{|T(n) - ET(n)| > t\} \leq \begin{cases} 2\exp(-ct^2/\log n), & d = 2 \\ 2\exp(-ct^2/n^{(d-2)/d}), & d \geq 3. \end{cases}$$

When seeking quick and easy deviation inequalities for Euclidean functionals, Azuma's inequality often suffices once bounds for the martingale differences $(d_i)_{i \geq 1}$ are in hand. For example, Talagrand (1991) uses a modification of this approach to find deviation bounds for the *directed* TSP, a topic addressed in Chapter 8.

However, the method of martingale differences, while general and simple, does not always yield optimal tail estimates, especially for problems in geometric probability. We will now use Azuma's inequality to develop improved estimates. These improved estimates, which are concentration inequalities obtained via isoperimetric methods, are crucial to our approach.

6.2.   The Rhee and Talagrand Concentration Inequalities

The seminal work of Talagrand (1995, 1996a) develops general isoperimetric inequalities for product measures which can be used to refine Theorem 6.1. Talagrand (1989,1994c) was motivated to develop his isoperimetric and concentration inequalities in order to investigate and settle various open problems related to sums of independent vector valued random variables. His results are a remarkable illustration of the power of abstract concentration of measure ideas; these ideas trace back to the work of V. Milman on the local theory of Banach spaces and in fact originate with Milman's proof of Dvoretzky's theorem on almost spherical sections of convex sets.

Loosely speaking, the concentration of measure phenomenon, which is at the heart of Talagrand's isoperimetric inequalities, says that if (X, d, μ) is a compact metric space, and the Borel set $B \subset X$ has a μ measure of at least one half, then "most" of the points in X are "close" to B. Talagrand's main contribution is to clarify in a mathematical sense the meaning of the words "most" and "close" for some natural families of spaces (X, d, μ). For example, if X is the Euclidean n-sphere S^n, d is the geodesic distance, μ is the normalized rotationally invariant measure, and B_ϵ represents the ϵ fattening of the Borel set B, $\mu(B) \geq 1/2$, then the words "most" and "close" mean that

$$(6.3) \qquad \mu(B_\epsilon) \geq 1 - (\frac{\pi}{8})^{1/2} \exp(-\epsilon^2(n-1)/2).$$

This inequality follows from Lévy's isoperimetric inequality; Talagrand's work has developed a new approach to providing isoperimetric inequalities which hold in more general settings. It may be shown that (6.3) implies that any "nice" function f on S^n, i.e, $f \in C(S^n)$, is close to a constant (its median value) everywhere except on a set whose measure is of the order $\exp(-\epsilon^2(n-1)/2)$. In other words, to use Milman's words, *a well-behaved function is "almost" a constant on "almost" all of the space.* This is the powerful concentration of measure phenomenon.

In a series of brilliant papers, Talagrand (1995,1996a,1996b) developed an entirely new set of isoperimetric inequalities which depends heavily upon novel ways to enlarge or fatten a set. Talagrand applies his inequalities to give a large number of applications in geometric probability, probability in Banach spaces, and percolation. We refer to Ledoux (1996) for a thorough and completely accessible treatment of isoperimetry.

Using his isoperimetric approach, Talagrand (1995) shows that when the Euclidean functional L is based on the TSP, MST, or Steiner MST problem, then $L(n) := L(U_1, ..., U_n)$ is concentrated around its mean in a remarkable way and actually exhibits sub-Gaussian behavior:

$$(6.4) \qquad P\{|L(n) - EL(n)| \geq t\} \leq C \exp(-t^2/C).$$

Notice that (6.4) goes well beyond the classical approach of Theorem 6.1, which depends upon the martingale difference method.

When L is the minimal matching functional S, then it is unclear whether the sub-Gaussian estimate (6.4) holds. This is due to the apparent lack of good regularity properties in minimal matching. Knowledge of the behavior of S on a set F does not in general tell much about the behavior of S on a modification of F. Indeed, small changes in F could lead to drastic changes in $S(F)$. Another approach, due to Rhee (1994b), gives the best currently available concentration estimates for S. The first of Rhee's results holds for the two dimensional case. As is customary by now, we let $S(n)$ denote $S(U_1, ..., U_n)$.

Theorem 6.2. *(concentration for minimal matching, $d = 2$) There is a constant $0 < C < \infty$ such that for all $t > 0$*

$$(6.5) \qquad P\{|S(n) - ES(n)| \geq t\} \leq C \exp(-t^2/C(\log n)^2).$$

A simple consequence of (6.5) is that the variance of $S(n)$ is at most $C(\log n)^2$. An apparently difficult question is whether the variance of $S(n)$ is bounded independently of n.

When the dimension d exceeds 2, Rhee (1993b) obtained sharper deviation estimates for $S(n)$. Her estimate is a consequence of a general deviation inequality for Euclidean functionals. We now discuss this important inequality, which forms the cornerstone of the entire theory.

We have seen that isoperimetric methods essentially show that "well-behaved" functions are close to their medians. In the setting of Euclidean functionals, it is reasonable to expect that if a functional L^p is smooth of order p (Hölder continuous) in the usual sense

$$|L^p(F \cup G) - L^p(F)| \leq (\text{card}\,G)^{(d-p)/d},$$

then it is well behaved and thus close to its mean. This is indeed the case as shown by the following general result, which applies to Euclidean functionals on sample points which need not be uniform.

Theorem 6.3. *(concentrations for Euclidean functionals, $d \geq 2$) Let X_i, $i \geq 1$, be independent random variables with values in $[0,1]^d$, $d \geq 2$. Let L^p, $0 < p < d$, be a Euclidean functional which is smooth of order p (3.8). Then there is a constant $C := C(L^p, d)$ such that for all $t > 0$ we have*

$$P\{|L^p(X_1, ..., X_n) - EL^p(X_1, ..., X_n)| > t\}$$

$$(6.6) \qquad \leq C \exp\left(\frac{-(t/C_3)^{2d/(d-p)}}{Cn}\right).$$

As we will soon see, Rhee (1993b) deduces (6.6) as a consequence of Azuma's inequality and an isoperimetric inequality for the Hamming distance. The upshot of (6.6) is that with high probability the functional $L^p(X_1, ..., X_n)$ and its mean $EL^p(X_1, ..., X_n)$ do not differ by more than $C(n \log n)^{(d-p)/2d}$. One of the most useful consequences of Rhee's concentration estimate (6.6) is that it reduces the problem of showing complete convergence of L^p to one of showing the convergence of the mean of L^p, a fact which we used heavily in the proof of the basic limit theorems of Chapter 4. We will also draw on this fact in the proof of our basic umbrella theorem of Chapter 7.

Corollary 6.4. (*convergence of means implies complete convergence*) *Let X_i, $i \geq 1$, be i.i.d. random variables with values in $[0,1]^d$ and law μ_X. Let L^p be a smooth Euclidean functional of order p, $0 < p < d$. If the mean of L^p converges in the sense that*

$$\lim_{n \to \infty} EL^p(X_1, ..., X_n)/n^{(d-p)/d} = \alpha(L_B^p, d, \mu_X),$$

then

(6.7) $$\lim_{n \to \infty} L^p(X_1, ..., X_n)/n^{(d-p)/d} = \alpha(L_B^p, d, \mu_X) \quad c.c.$$

Proof. For all $\epsilon > 0$ we have by (6.6)

$$\sum_{n=1}^{\infty} P\left\{ \left| \frac{L^p(X_1, ..., X_n) - EL^p(X_1, ..., X_n)}{n^{(d-p)/d}} \right| > \epsilon \right\}$$

$$\leq C \sum_{n=1}^{\infty} \exp\left(-(\frac{\epsilon}{C_3})^{2d/(d-p)} \frac{n}{C} \right).$$

Thus $\left| \frac{L^p(X_1, ..., X_n) - EL^p(X_1, ..., X_n)}{n^{(d-p)/d}} \right|$ converges completely to zero and the proof is complete. $\square$

Another consequence of Theorem 6.3 is that, together with the rate results of Chapter 5, we can deduce high probability rate results for the functional $L^p(U_1, ..., U_n)$. Combining Theorem 6.3 and (5.3), for example, gives for all $d \geq 3$ and $0 < p < d$ the high probability estimate

$$|L^p(U_1, ..., U_n) - \alpha(L_B^p, d)n^{(d-p)/d}| \leq C \left(n^{(d-p-1)/d} \vee (n \log n)^{(d-p)/2d} \right).$$

6.3. Isoperimetry

In this section we describe the isoperimetric methods which are behind the proof of Theorem 6.3 and which form one of the central themes of this monograph. The approach is guided by Steele (1997).

The path to estimates of the type (6.3) and (6.4) proceeds via isoperimetric inequalities. While there are many such inequalities we will focus on that which involves the Hamming distance H on n-fold product spaces Ω^n.

The Hamming distance H on Ω^n measures the distance between x and y by the number of coordinates in which x and y disagree, that is

$$H(x,y) := \operatorname{card}\{i : x_i \neq y_i\}.$$

To formulate our isoperimetric inequality, we will take $\Omega := [0,1]^d$ and μ a measure on $[0,1]^d$; μ^n denotes the product measure on the product space $([0,1]^d)^n$. Given $A \subset \Omega^n$ and $y \in \Omega^n$, we define the Hamming distance between y and A by

$$\phi_A(y) := \min\{H(x,y) : x \in A\}.$$

We will assume that A satisfies $\mu^n(A) \geq 1/2$.

Changing one of the n coordinates of y produces a change of at most 1 in $H(x,y)$ and therefore the martingale differences d_i, $1 \leq i \leq n$, appearing in the martingale difference representation of $\phi_A(y)$ are bounded by 1. By Azuma's inequality it follows that

$$(6.8) \qquad \mu^n(y : |\phi_A(y) - \alpha| \geq t) \leq 2\exp(-t^2/2n),$$

where $\alpha := \int \phi_A(y) d\mu^n$. Using $\phi_A(y) = 0$ for $y \in A$, we see that the left side of (6.8) is at least as large as $\mu^n(A)$ when $t = \alpha$. Since $\mu^n(A) \geq 1/2$ it follows that $1/2 \leq 2\exp(-\alpha^2/2n)$ or $\alpha \leq (2n \log 4)^{1/2}$. By (6.8) it follows that

$$\mu^n(\phi_A(y) \geq t + (2n \log 4)^{1/2}) \leq 2\exp(-t^2/2n)$$

and therefore for $t \geq 2(2n \log 4)^{1/2}$ we obtain

$$\mu^n(\phi_A(y) \geq t) = \mu^n\left(\phi_A(y) \geq t - (2n \log 4)^{1/2} + (2n \log 4)^{1/2}\right)$$

$$(6.9) \qquad\qquad\qquad \leq 2\exp(-t^2/8n)$$

since $(t - (2n \log 4)^{1/2})^2 \geq (t/2)^2$ when t is in the range $2(2n \log 4)^{1/2} \leq t < \infty$. In the range $0 \leq t \leq 2(2n \log 4)^{1/2}$ we have $2\exp(-t^2/8n) \geq 1/2$ and so (6.9) holds for all $0 \leq t < \infty$ provided that the coefficient 2 is replaced by 4.

We have thus proved

Proposition 6.5. *(isoperimetry) If $A \subset \Omega^n$ satisfies $\mu^n(A) \geq 1/2$ then*

$$(6.10) \qquad \mu^n(\{y \in \Omega^n : \phi_A(y) \geq t\}) \leq 4\exp(-t^2/8n).$$

Proposition 6.5 is an isoperimetric inequality for the Hamming distance H. This inequality goes back to Milman and Schechtman (1986) and its relevance to optimization problems was recognized by Rhee (1993b, Proposition 3). For further refinements of (6.10) with sharper constants we refer to Talagrand (1995). To understand how Proposition 6.5 captures isoperimetry and to see its relation to the classic inequality (6.3), consider the following. As is customary in isoperimetry, we define the t-enlargement of $A \subset \Omega^n$:

$$A_t := \{x \in \Omega^n : \exists\, y \in A \quad \text{such that} \quad H(x,y) \leq t\}.$$

Proposition 6.5 now says that if $\mu^n(A) \geq 1/2$ then

$$(6.11) \qquad \mu^n(A_t) \geq 1 - 4\exp(-t^2/8n).$$

If $t \geq Cn^{1/2}$ then (6.11) implies that the measure of the fattened set A_t is almost 1. We will use both (6.10) and (6.11) in the sequel.

We now use Proposition 6.5 to prove Theorem 6.3.

Proof of Theorem 6.3. We will closely follow Steele's (1997) exposition of Rhee's (1993b) proof. Let $Z := L^p(X_1, ..., X_n)$ and let $M := M(n)$ denote a median of Z. If $A := \{x = (x_1, ..., x_n) \in ([0,1]^d)^n : L^p(x) \leq M\}$ then for each $y = (y_1, ..., y_n) \in ([0,1]^d)^n$ there is an $x \in A$ such that $H(x,y) := \phi_A(y)$. Given y we let $F := F(y,x)$ denote the coordinates of x agreeing with those in y, i.e.

$$F := \{x_i, 1 \leq i \leq n : x_i = y_i\}.$$

We let $G := G(y,x)$ denote the remaining coordinates, i.e., $G := \{x_i, 1 \leq i \leq n : x_i \neq y_i\}$. Note that $\operatorname{card} G = \phi_A(y)$.

By the assumed smoothness, we have that both $|L^p(y) - L^p(F)|$ and $|L^p(F) - L^p(x)|$ are bounded by $C_3(\operatorname{card} G)^{(d-p)/d}$. Writing

$$L^p(y) \leq |L^p(y) - L^p(F)| + |L^p(F) - L^p(x)| + L^p(x),$$

we obtain

$$L^p(y) \leq 2C_3\phi_A(y)^{(d-p)/d} + M$$

since $L^p(x) \leq M$.

Letting μ be the law of X_1 we obtain by Proposition 6.5

$$P\{Z \geq M + t\} = \mu^n(\{y \in \Omega^n : L^p(y) \geq M + t\})$$

$$\leq \mu^n(\phi_A(y)^{(d-p)/d} \geq \frac{t}{2C_3})$$

$$\leq 4\exp\left(\frac{-t^{2d/d-p}}{8n(2C_3)^{2d/d-p}}\right).$$

Similar arguments show that $P\{Z \leq M - t\}$ satisfies the same bound and therefore

$$(6.12) \qquad P\{|Z - M| \geq t\} \leq 8\exp\left(\frac{-t^{2d/d-p}}{8n(2C_3)^{2d/d-p}}\right).$$

Integrating this tail bound it follows that $E|Z - M| \leq Cn^{(d-p)/2d}$ and thus $|EZ - M| \leq Cn^{(d-p)/2d}$. Consequently, by (6.12) we have

$$(6.13) \qquad P\{|Z - EZ| \geq t + Cn^{(d-p)/2d}\} \leq 8\exp\left(\frac{-t^{2d/d-p}}{8n(2C_3)^{2d/d-p}}\right).$$

We convert (6.13) into a tail bound for $|Z - EZ|$. When $t \geq 2Cn^{(d-p)/2d}$ we write $t = t - Cn^{(d-p)/2d} + Cn^{(d-p)/2d}$ and obtain

$$(6.14) \qquad P\{|Z - EZ| \geq t\} \leq 8\exp\left(\frac{-t^{2d/d-p}}{8n(4C_3)^{2d/d-p}}\right),$$

where we use $(t - Cn^{(d-p)/2d})^{2d/(d-p)} \geq (t/2)^{2d/d-p}$ for $t \geq 2Cn^{(d-p)/2d}$. In the range $0 \leq t \leq 2Cn^{(d-p)/2d}$ the right side of (6.14) is bigger than a positive constant and therefore $P\{|Z - EZ| \geq t\}$ is bounded by the right side of (6.14) divided by this constant. This is precisely the desired inequality (6.6). $\square$

6.4. Isoperimetry and the Power-Weighted MST

In this section we illustrate one of the many applications of isoperimetric methods. We show that isoperimetry may be used to strengthen Theorem 4.6 to a complete convergence result. Similar methods apply to Theorem 4.7 as well. Specifically we will prove:

Theorem 6.6. Let U_i, $i \geq 1$, be i.i.d. with the uniform distribution on $[0,1]^d$. Then

$$(6.15) \qquad \lim_{n\to\infty} M^d(U_1, ..., U_n) = \alpha(M_B^d, d) \quad c.c.$$

Proof. The proof uses a simple variant of Proposition 6.5 to derive a concentration inequality for the power-weighted MST about its median. As before we set $\Omega = [0,1]^d$; we let μ^n denote the uniform measure on Ω^n.

We let $M := M(n)$ denote a median of $M^d(U_1, ..., U_n)$ and we let $A \subset \Omega^n$ consist of those n-tuples $x := \{x_1, ..., x_n\} \in \Omega^n$ for which

$$M^d(x) := M^d(x_1, ..., x_n) \geq M.$$

Recall that for all $t > 0$ the t-enlargement of A is

$$A_t := \{x \in \Omega^n : \exists\, y \in A \text{ such that } H(x,y) \leq t\}.$$

We define two more sets. Let $B \subset \Omega^n$ consist of those points $x := (x_1, ..., x_n)$ such that the edges in the MST graph on $\{x_i\}_{i=1}^n$ have length at most $C(\log n/n)^{1/d}$. Let

$$D := \{x = (x_1, ..., x_n) \in \Omega^n : \max_{j \leq n} d(g_j, \{x_i\}_{i=1}^n) \leq C(\log n/n)^{1/d}\}$$

where $\{g_j\}_{j=1}^n$ denotes grid points in $[0,1]^d$ and where $d(x, F)$ denotes the distance between the point x and the set F.

Since $\mu^n(A) \geq 1/2$ and since B and D are high probability sets for μ^n, we easily have $\mu^n(A \cap B \cap D) \geq 1/3$. By a simple variant of Proposition 6.5 it follows that if $\mu^n(A \cap B \cap D) \geq 1/3$, then

$$\mu^n(\{y \in \Omega^n : \phi_{A \cap B \cap D}(y) \geq t\}) \leq 6\exp(-t^2/8n).$$

Therefore the enlarged set $(A \cap B \cap D)_{tn^{1/2}}$ occurs with high probability:

$$\mu^n((A \cap B \cap D)^c_{tn^{1/2}}) \leq 6\exp(-t^2/8).$$

Now define $E := B \cap D \cap (A \cap B \cap D)_{tn^{1/2}}$ and note for all $\beta > 1$ that a suitable choice for C yields the high probability bound:

$$\mu^n(E^c) \leq n^{-\beta} + 6\exp(-t^2/8).$$

If $x = (x_1, ..., x_n) \in E$ then there is a point $y = y(x) = (y_1, ..., y_n)$ in $A \cap B \cap D$ such that the following conditions hold: $H(x, y) \leq tn^{1/2}$, y is close to x in the sense that $\max_{i \leq n} d(x_i, \{y_j\}_{j=1}^n) \leq C(\log n/n)^{1/d}$ and $\max_{i \leq n} d(y_i, \{x_j\}_{j=1}^n) \leq C(\log n/n)^{1/d}$, and the edges in the graph of the minimal spanning tree on y have length bounded by $C(\log n/n)^{1/d}$.

We now claim that for this choice of y we have

$$|M^d(x) - M^d(y)| \leq Ct \log n/n^{1/2}.$$

Indeed, to see that

$$M^d(x) \leq M^d(y) + Ct \log n/n^{1/2},$$

consider the coordinates of x which differ from the coordinates in y. Join these coordinates to points in $\{y_i\}_{i=1}^n$ at a cost of at most $H(x, y) \cdot \log n/n$. This produces a spanning graph G on the union $\{x_i\}_{i=1}^n \cup \{y_i\}_{i=1}^n$. To obtain a spanning graph on just $\{x_i\}_{i=1}^n$ we may modify the edges in G which are linked to those y coordinates which do not appear in x. Using the approach of Lemma 4.8 we see that this may be done at a cost of at most $H(x, y) \cdot \log n/n$, thus showing the desired inequality. The proof of the reverse inequality

$$M^d(y) \leq M^d(x) + Ct \log n/n^{1/2}$$

holds for the same reasons and this proves the stated claim.

Therefore if $x \in E$ we have

$$M^d(x) \geq M^d(y) - |M^d(x) - M^d(y)| \geq M - Ct \log n/n^{1/2}.$$

Thus for all $0 \leq t \leq n^{1/2}/2$ it follows that

$$P\{M^d(U_1, ..., U_n) \leq M - Ct \cdot \log n/n^{1/2}\}$$
$$\leq \mu^n(E^c)$$
$$\leq n^{-\beta} + 6\exp(-t^2/8).$$

Using a similar argument for the reverse inequality

$$P\{M^d(U_1, ..., U_n) \geq M + Ct \cdot \log n/n^{1/2}\}$$

we obtain for all $0 \leq t \leq n^{1/2}/2$

$$P\{|M^d(U_1, ..., U_n) - M| \geq Ct \cdot \log n/n^{1/2}\} \leq 2n^{-\beta} + 12 \exp(-t^2/8).$$

Setting $t := \epsilon \cdot n^{1/2}/\log n$, where $\epsilon > 0$ is arbitrary but fixed, yields the concentration inequality

(6.16) $$P\{|M^d(U_1, ..., U_n) - M| \geq \epsilon\} \leq 2n^{-\beta} + 12\exp(-\epsilon^2 n/8(\log n)^2).$$

The arbitrariness of ϵ and the Borel-Cantelli lemma imply that

$$\lim_{n \to \infty} |M^d(U_1, ..., U_n) - M| = 0 \quad c.c.$$

Integrating (6.16) also shows that $\lim_{n \to \infty} E|M^d(U_1, ..., U_n) - M| = 0$ and thus

$$\lim_{n \to \infty} |EM^d(U_1, ..., U_n) - M| = 0.$$

Since Theorem 4.6 gives

$$\lim_{n \to \infty} |EM^d(U_1, ..., U_n) - \alpha(M_B^d, d)| = 0$$

it follows from the triangle inequality that

$$\lim_{n \to \infty} M^d(U_1, ..., U_n) = \alpha(M_B^d, d) \quad c.c.$$

as desired. □

6.5. Large Deviations

In this section we use the deviation estimate (6.4) and the superadditivity of the boundary TSP functional T_B to arrive at a large deviation principle for T_B. This principle will show that the sub-Gaussian tail behavior of (6.4) cannot be sharpened and in this way it further supports the conjecture that the TSP functional has an underlying Gaussian (normal) structure. The approach originates in discussions with Amir Dembo and Ofer Zeitouni and it is a pleasure to thank them for their ideas.

For all $1 \leq i \leq 4$, let U_{ij}, $j \geq 1$, be i.i.d. uniform random variables in the subsquare Q_i and let $N_i(n)$ be an independent Poisson random variable with parameter n. The superadditivity of T_B gives

$$T_B(U_{11}, ..., U_{1N_1(n)}, U_{21}, ..., U_{2N_2(n)}, U_{31}, ..., U_{3N_3(n)}, U_{41}, ..., U_{4N_4(n)})$$

$$\geq \sum_{i=1}^{4} T_B(U_{i1}, ..., U_{iN_i(n)}, Q_i).$$

Write $T_B(k) := T_B(U_1, ..., U_{N(k)})$ where U_i, $i \geq 1$, are i.i.d. uniform random variables with values in $[0, 1]^d$ and $N(k)$ is an independent Poisson random variable with parameter k. Then the left side of the above is equal in distribution to $T_B(4n)$ and by scaling the right side is equal in distribution to $\frac{1}{2} \sum_{i=1}^{4} T_B^{(i)}(n)$, where $T_B^{(i)}(n)$ are independent copies of $T_B(n)$. It follows that for all $t > 0$

$$P\{T_B(4n) > t\} \geq P\{\frac{1}{2} \sum_{i=1}^{4} T_B^{(i)}(n) > t\}$$

and thus

$$P\{T_B(4n)/(4n)^{1/2} > t\} \geq P\left\{ \frac{\sum_{i=1}^{4} T_B^{(i)}(n)}{4n^{1/2}} > t \right\}$$

$$\geq \left(P\{T_B(n)/n^{1/2} > t\} \right)^4.$$

More generally, for all positive integers m and n we have

$$P\left\{ \frac{T_B(4^m n)}{(4^m n)^{1/2}} > t \right\} \geq \left(P\{\frac{T_B(n)}{n^{1/2}} > t\} \right)^{4^m}.$$

If we set for all $n \in \mathbb{N}$

$$\phi(n) := -\log P\{\frac{T_B(n)}{n^{1/2}} > t\}$$

then the above relation tells us that $\phi(4^m n) \leq 4^m \phi(n)$. Homogenizing, we arrive at

$$\frac{\phi(4^m n)}{4^m n} \leq \frac{\phi(n)}{n}$$

and more generally, letting $n = 4^k$ we have

(6.17)
$$\frac{\phi(4^{m+k})}{4^{m+k}} \leq \frac{\phi(4^k)}{4^k}$$

for all positive integers m and k. If we now set $\alpha(j) := \phi(4^j)/4^j$ then it follows that $\alpha(j)$ is decreasing by (6.17). On the other hand, it follows by the sub-Gaussian estimate (6.4) that if $t > C_2 := \limsup_{j \to \infty} \frac{ET_B(4^j)}{2^j}$ then for j large

$$
\begin{aligned}
P\{T_B(4^j)/2^j > t\} &\leq P\{|T_B(4^j) - ET_B(4^j)| > t \cdot 2^j - ET_B(4^j)\} \\
&\leq P\{|T_B(4^j) - ET_B(4^j)| > (t - C_2) \cdot 2^j\} \\
&\leq C \cdot \exp(-(t - C_2)^2 \cdot 4^j/C),
\end{aligned}
$$

and therefore for j large, $\alpha(j)$ is bounded below by a positive constant depending only on t. Since $\alpha(j)$ is decreasing, we get the limit

$$
\lim_{j \to \infty} \alpha(j) = C(t),
$$

where $C(t)$ depends only on t. We have thus shown the following large deviation principle:

Theorem 6.7. *For* $t > C_2 := \limsup_{j \to \infty} \frac{ET_B(4^j)}{2^j}$ *there is a positive finite constant* $C(t)$ *such that*

(6.18) $$\lim_{j \to \infty} \frac{-\log P\{T_B(4^j)/2^j > t\}}{4^j} = C(t).$$

The importance of (6.18) is that it essentially shows that the sub-Gaussian tail behavior (6.4) cannot be improved and that it is of the correct order of magnitude. Theorem 6.7 complements the work of Rhee (1991), who obtains lower bounds for the tails of the TSP. Writing, as usual, $T(n)$ for $T(U_1, ..., U_n)$, Rhee (1991) proves that there is a universal positive constant C such that for all $0 < t \leq C^{-1} n^{1/2}$ we have

(6.19) $$P\{T(n) \leq ET(n) - t\} \geq C^{-1} \exp(-t^2 C).$$

This shows that the sub-Gaussian estimate (6.4) is "sharp" in the setting of the TSP.

Still, Theorem 6.7 raises some additional questions. For example, can we prove an analogous result for the random variables $T_B(n)$ when n ranges over all integers and not just powers of 4? Is it possible to find the analog of (6.18) for the standard TSP functional T?

Since the sub-Gaussian estimate (6.4) also holds for the MST functional, it is clear that Theorem 6.7 also holds for the boundary MST functional M_B. However, it is not yet clear that it holds for the boundary minimal matching functional S_B.

We anticipate that Theorem 6.7 can be turned into a genuine large deviation principle. Given a Euclidean functional L on $\mathbb{R}^2$ of order 1, let $L(n) := L(U_1, ..., U_n)$ and let $A := A(L) := \sup_n L(n)/n^{1/2}$. We anticipate that the following large deviation principle holds:

Conjecture 6.8. *Let L be a subadditive Euclidean functional on $\mathbb{R}^2$ which is smooth of order 1 and pointwise close to the superadditive boundary functional L_B. There is a rate function $I(x)$ such that for all closed sets $F \subset [0, A]$ we have*

$$(6.20) \qquad \limsup_{n \to \infty} \frac{\log P\{L(n)/n^{1/2} \in F\}}{n} \leq - \inf_{x \in F} I(x),$$

and for all open sets $O \subset [0, A]$ we have

$$(6.21) \qquad \liminf_{n \to \infty} \frac{\log P\{L(n)/n^{1/2} \in O\}}{n} \geq - \inf_{x \in O} I(x).$$

Remark.

It is *a priori* not clear whether (6.20) and (6.21) give the correct rates of convergence. However, a little reflection shows that the rates are indeed the right ones, at least when L is the TSP functional T. Indeed, by the sub-Gaussian estimate (6.4) we obtain for all $t > \beta := \sup_n ET(n)/n^{1/2}$ that

$$P\{T(4^k)/2^k \geq t\} \leq C \exp\left(-(t - \beta)^2 4^k / C\right).$$

Therefore we have

$$\limsup_{k \to \infty} \frac{\log P\{T(4^k)/2^k \geq t\}}{4^k} \leq -(t - \beta)^2 / C < 0.$$

By Rhee's lower bounds (6.19) we similarly obtain for all $t < \beta$ the lower estimate

$$\liminf_{k \to \infty} \frac{\log P\{T(4^k)/2^k \leq t\}}{4^k} \geq -C(\beta - t)^2 > -\infty.$$

Thus the rates given by (6.20) and (6.21) are of the right order.

Notes and References

1. Section 6.3 follows closely the exposition of Steele (1997) and Rhee (1993b), who treat the case $p = 1$. Talagrand (1996c) suggested the use of isoperimetry to obtain the complete convergence of the power-weighted MST. For extensions of Theorem 6.6 to non-uniform random variables see Yukich (1997b).

2. It would be useful to obtain a concentration estimate for subadditive Euclidean functionals L^p when $p \geq d$. This would extend Theorem 6.3.

3. Azuma's inequality is a standard tool and is useful in many contexts. Our treatment is based on Ledoux and Talagrand (1991).

7. UMBRELLA THEOREMS FOR EUCLIDEAN

FUNCTIONALS

7.1. The Basic Umbrella Theorem

In 1959 Beardwood, Halton, and Hammersley proved their celebrated result describing the asymptotic length of the shortest tour on a random sample. They showed that the shortest tour $T(X_1, ..., X_n)$ through i.i.d. random variables X_i, $i \geq 1$, with values in $[0,1]^d$ satisfies

$$(7.1) \qquad \lim_{n \to \infty} T(X_1, ..., X_n)/n^{(d-1)/d} = \alpha(d) \int_{[0,1]^d} f(x)^{(d-1)/d} dx \quad a.s.,$$

where f is the density of the absolutely continuous part of the law of X_1 and $\alpha(d)$ is a positive constant. This seminal limit law extends the limit law encountered in Chapter 4 to the non-uniform case.

Chapter 4 established (7.1) for the special case of uniformly distributed X_i, $i \geq 1$. One expects that the general case follows by suitably approximating general random variables by linear combinations of uniform random variables and then taking limits. Despite the lack of a suitable convergence theorem, this "standard approach" can be made to work, but only by calling upon the tools and methods developed in previous chapters. This is where isoperimetry and the complementary notions of subadditive and superadditive functionals play an indispensable role.

The asymptotics (7.1) for the TSP provide enticing evidence that the basic limit Theorems 4.1, 4.3, and 4.5 may be extended to the non-uniform setting. Under what conditions can the uniform random variables $U_1, ..., U_n$ of Theorems 4.1 and 4.3 be replaced by non-uniform random variables $X_1, ..., X_n$? Following Steele (1981a, 1988), and Rhee (1993b) we reformulate this question and ask: can the asymptotics (7.1) be usefully generalized in terms of an umbrella result which includes the solutions to a broad range of problems in Euclidean optimization? Such an umbrella result would ideally cover those Euclidean functionals lacking monotonicity as well as those having power-weighted edges.

The following general result, together with the basic limit Theorem 4.1, lie at the heart of this monograph. The constant $\alpha(L_B^p, d)$ is of course the same constant (4.5) appearing earlier.

Theorem 7.1. (umbrella theorem for Euclidean functionals on compact sets) Let L^p and L_B^p be smooth subadditive and superadditive Euclidean functionals of order p, respectively. Assume that L^p and L_B^p are close in mean (3.15). Let $(X_i)_{i \geq 1}$ be i.i.d. random variables with values in $[0,1]^d$, $d \geq 2$. If $1 \leq p < d$ then

$$(7.2) \qquad \lim_{n \to \infty} L^p(X_1, ..., X_n)/n^{(d-p)/d} = \alpha(L_B^p, d) \int_{[0,1]^d} f(x)^{(d-p)/d} dx \quad c.c.,$$

where f is the density of the absolutely continuous part of the law of X_1.

We will shortly see that Theorem 7.1 is truly an umbrella result in the sense that it captures c.c. asymptotics for a wide range of problems in geometric probability, including those in combinatorial optimization, operations research, and computational geometry. Notice that Theorem 7.1 provides the asymptotics for the archetypical problems (TSP, MST, and minimal matching) considered up to now. Indeed, recalling Lemma 3.5 we know that the TSP, MST, and minimal matching functionals T^p, M^p, and S^p, respectively, are smooth Euclidean functionals of order p. By Lemma 3.10, these functionals are close in mean to their respective boundary functionals T_B^p, M_B^p, and S_B^p. Therefore we have shown:

Corollary 7.2. *The Euclidean functionals T^p, M^p, and S^p, $1 \leq p < d$, and their corresponding boundary functionals all exhibit the asymptotic behavior (7.2).*

Remarks.

(i) Corollary 7.2 is not exhaustive. We will show that other problems in geometric probability satisfy the hypotheses of Theorem 7.1. This will include the semi-matching problem and the k nearest neighbors problem (Chapter 8), the minimal triangulation problem (Chapter 9), and geometric location problems (Chapter 10).

(ii) Returning to (7.2), we clearly have $L^p(X_1, ..., X_n) = o(n^{(d-p)/d})$ a.s. when X_1 has a singular distribution. On the other hand, Jensen's inequality shows that the right side of (7.2) is largest when the density $f(x)$ equals $1_{[0,1]^d}(x)$, that is when the X_i, $i \geq 1$, have the uniform distribution on the unit cube. Thus Euclidean functionals L^p assume their largest value (in the asymptotic sense) when the underlying sample is uniformly distributed.

(iii) The TSP, MST, and minimal matching functionals may be defined on the torus T, defined as the unit cube equipped with the flat metric or Euclidean d-torus metric. The weight of an edge (x_i, x_j) is now $\|(x_i - x_j)(\mathrm{mod} 1)^d\|$. It is easily checked that the resulting functionals, which we call T_T^p, M_T^p, and S_T^p, respectively, are sandwiched between the boundary functional and the standard functional. In other words, $T_B^p \leq T_T^p \leq T^p$ and similarly for M_T and S_T. By Corollary 7.2 these functionals all exhibit the asymptotic behavior (7.2). Jaillet (1993b), using different methods, was the first to observe that the torus versions of our archetypical functionals satisfy the limit law (7.2).

(iv) For Euclidean functionals L^p of the generality described in Theorem 7.1, the use of boundary functionals L_B^p is critical in order to obtain the asymptotics (7.2). In the lucky event that L^p satisfies the extra side condition of monotonicity (that is $L^p(F) \leq L^p(F \cup \{x\})$ for all sets F and singletons x) then it is possible to deduce (7.2) in a relatively straightforward way. Indeed, we first prove (7.2) when the law

μ of X_1 is absolutely continuous with a step function density, then when μ is the sum of a step function density and a singular part, and finally when μ is a mixture of absolutely continuous and singular laws (see Steele (1997)). For the last step, coupling methods are useful. However, without monotonicity of L^p, considerable effort is required to handle distributions having a singular part. In particular, it is especially difficult to obtain the lower bound implicit in (7.2). To see that there are real difficulties in proving the lower bound, the reader may try to prove (7.2) for the MST functional. Steele (1988) addresses and overcomes these difficulties.

However, with the use of boundary functionals the proof of (7.2) becomes much easier. Indeed, once we use the subadditivity of L^p to prove the upper bound in (7.2) then exactly *the same methods* may be used to show that the superadditive functional L^p_B satisfies the lower bound in (7.2). Since L^p and L^p_B are close in mean, (7.2) follows. This is a brief outline of the main idea of the proof of (7.2). The details and the complete proof of (7.2) are in the next section.

7.2. Proof of the Basic Umbrella Theorem

The proof of (7.2) depends upon two observations which greatly simplify the analysis. The first is that by Corollary 6.4, it is enough to show that (7.2) holds in expectation, namely it suffices to show

$$(7.3) \qquad \lim_{n \to \infty} EL^p(X_1, ..., X_n)/n^{(d-p)/d} = \alpha(L^p_B, d) \int_{[0,1]^d} f(x)^{(d-p)/d} dx.$$

The limit (7.3) is a statement about a sequence of scalars and is thus easier to prove than the limit (7.2).

The second observation is that in the presence of the assumed smoothness of L^p, it is enough to establish (7.3) for a special class of distributions which we call *blocked distributions*. These are distributions μ on $[0,1]^d$ with the form $\phi(x)dx + \mu_s$, where $\phi(x)$ is a simple non-negative function of the form $\sum_{i=1}^{m^d} \alpha_i 1_{Q_i}$, the measure μ_s is purely singular, m is a power of 2, and Q_i, $i \geq 1$, are the usual subcubes. More precisely, we have the following lemma which is due to Steele (1988).

Lemma 7.3. *(reduction to blocked distributions) Let L^p be a smooth subadditive Euclidean functional and suppose that for every sequence of i.i.d. random variables $(X_i)_{i \geq 1}$ distributed with a blocked distribution $\mu := \phi(x)dx + \mu_s$, we have that*

$$(7.4) \qquad \lim_{n \to \infty} EL^p(X_1, ..., X_n)/n^{(d-p)/d} = \alpha(L^p_B, d) \int_{[0,1]^d} \phi(x)^{(d-p)/d} dx.$$

We then have that

$$(7.5) \qquad \lim_{n \to \infty} EL^p(Y_1, ..., Y_n)/n^{(d-p)/d} = \alpha(L^p_B, d) \int_{[0,1]^d} f(x)^{(d-p)/d} dx$$

whenever $(Y_i)_{i \geq 1}$ are independent and identically distributed with respect to any probability measure on $[0,1]^d$ with an absolutely continuous part given by $f(x)dx$.

Proof. The proof evolves from a coupling argument which is discussed and proved by Steele (1988, Theorem 3). Assume that the distribution of Y has the form $\mu_Y := f(x)dx + \mu_s$, where μ_s is singular. For all $\epsilon > 0$ we may find a blocked approximation to μ_Y of the form $\mu_X := \phi(x)dx + \mu_s$ where $\phi := \phi_\epsilon$ approximates f in the L^1 sense:

(7.6)
$$\int_{[0,1]^d} |\phi(x) - f(x)|dx < \epsilon.$$

By standard coupling arguments there is a joint distribution for the pair of random variables (X, Y) such that $P\{X \neq Y\} \leq 2\epsilon$. Thus it follows that

$$|EL^p(X_1, ..., X_n) - EL^p(Y_1, ..., Y_n)| \leq CE(\text{card}\{i \leq n : X_i \neq Y_i\}^{(d-p)/d})$$

(7.7)
$$\leq C(\epsilon n)^{(d-p)/d}.$$

Thus by (7.4) we obtain

(7.8)
$$\lim_{n\to\infty} \left| \frac{EL^p(Y_1, ..., Y_n)}{n^{(d-p)/d}} - \alpha(L_B^p, d) \int_{[0,1]^d} \phi(x)^{(d-p)/d}dx \right| \leq C\epsilon^{(d-p)/d}.$$

For all a, $b \geq 0$ we have

$$|a^{(d-p)/d} - b^{(d-p)/d}| \leq |a - b|^{(d-p)/d}$$

and therefore by (7.6)

$$\left| \int f(x)^{(d-p)/d}dx - \int \phi(x)^{(d-p)/d}dx \right| \leq \int |f(x) - \phi(x)|^{(d-p)/d}dx$$

(7.9)
$$< \epsilon^{(d-p)/d}.$$

Combining (7.8) and (7.9) and letting ϵ tend to zero gives the result (7.5) as desired. $\square$

We now prove (7.3) for the blocked distributions

$$\mu(x) := \sum_{i=1}^{m^d} \alpha_i 1_{Q_i}(x)dx + \mu_s$$

and we set $\phi(x) := \sum_{i=1}^{m^d} \alpha_i 1_{Q_i}(x)$. We will follow an approach similar to that of Steele (1981a,1988) and Redmond and Yukich (1994). Fix $\epsilon > 0$ and assume without loss of generality that $m^{-1} < \epsilon$. We will assume that m is a power of 2 so that we can apply geometric subadditivity (3.5). Let E denote the singular support of μ and let λ denote Lebesgue measure on the cube.

We may assume that m is chosen so that:

(1) $E \subset A \cup B$, where A and B are disjoint, $\lambda(A) = 0$ and $\mu(A) \leq \epsilon$, and

(2) $B := \bigcup_{i \in J} Q_i$ for some $J \subset I := \{1, ..., m^d\}$ and moreover $\lambda(B) \leq \epsilon$.

By smoothness (3.8), property (1), Jensen's inequality, and geometric subadditivity (3.5) we have

$$
\begin{aligned}
EL^p(X_1, ..., X_n) &\leq EL^p(\{X_1, ..., X_n\} - A) + C_3(\epsilon n)^{(d-p)/d} \\
&\leq \sum_{i \in I - J} EL^p(\{X_1, ..., X_n\} - A \cap Q_i, Q_i) + \\
&\quad + \sum_{i \in J} EL^p(\{X_1, ..., X_n\} - A \cap Q_i, Q_i) + \\
&\quad + C_1 m^{d-p} + C_3(\epsilon n)^{(d-p)/d}.
\end{aligned}
$$

(7.10)

Letting $(U_k)_{k \geq 1}$ be i.i.d. with the uniform distribution on $[0, 1]^d$ it follows by smoothness and homogeneity that the first sum in (7.10) is bounded by

$$
m^{-p} \sum_{i \in I - J} \left(EL^p((U_k)_{k=1}^{[\alpha_i m^{-d} n]}) + C_3 \left(E|B(n, \alpha_i m^{-d}) - [n\alpha_i m^{-d}]|)^{(d-p)/d} \right) \right)
$$

since for $i \in I - J$ the number of points not in A and in the subcube Q_i is a binomial random variable $B(n, \alpha_i m^{-d})$ with parameters n and $\alpha_i m^{-d}$. By Jensen's inequality, the above is clearly bounded by

$$
m^{-p} \sum_{i \in I - J} \left(EL^p((U_k)_{k=1}^{[\alpha_i m^{-d} n]}) + C(m) n^{(d-p)/2d} \right),
$$

where $C(m)$ is a constant depending only on d, m, and p. We now consider the second sum in (7.10). The expected number of points in $Q_i - A$ is at most $n\mu(Q_i)$. By Jensen's inequality and the growth bounds of Lemma 3.3 the second sum in (7.10) is bounded by

$$
\begin{aligned}
C_2 \sum_{i \in J} m^{-p}(n\mu(Q_i))^{(d-p)/d} &= C_2 n^{(d-p)/d} \sum_{i \in J} (m^{-d})^{p/d} \mu(Q_i)^{(d-p)/d} \\
&\leq C_2 n^{(d-p)/d} \left(\sum_{i \in J} m^{-d} \right)^{p/d} \\
&= C_2 n^{(d-p)/d} (\lambda(B))^{p/d} \\
&\leq C_2 \, \epsilon^{p/d} n^{(d-p)/d}
\end{aligned}
$$

by Hölder's inequality and the estimate $\lambda(B) \leq \epsilon$.

Combining the above estimates and dividing (7.10) by $n^{(d-p)/d}$ we get

$$EL^p(X_1, ..., X_n)/n^{(d-p)/d}$$
$$\leq \sum_{i \in I-J} m^{-p}([\alpha_i m^{-d} n]/n)^{(d-p)/d} \cdot EL^p\left((U_k)_{k=1}^{[\alpha_i m^{-d} n]}\right)/[\alpha_i m^{-d} n]^{(d-p)/d} +$$
$$+ C(m)n^{(p-d)/2d} + C_2 \epsilon^{p/d} + C_1 m^{(d-p)}/n^{(d-p)/d} + C_3 \epsilon^{(d-p)/d}.$$

Since the right side involves the L^p functional over a sequence of uniform random variables, we may evaluate this by applying the basic limit theorems of Chapter 4. Therefore, letting n tend to infinity, applying Theorem 4.1, and recalling that $1 \leq p < d$, we obtain

$$\limsup_{n \to \infty} EL^p(X_1, ..., X_n)/n^{(d-p)/d}$$
$$\leq \sum_{i \in I-J} \alpha_i^{(d-p)/d} m^{-d} \alpha(L_B^p, d) + C_2 \epsilon^{p/d} + C_3 \epsilon^{(d-p)/d}$$
$$= \alpha(L_B^p, d) \int_{\cup_{i \in I-J} Q_i} \phi(x)^{(d-p)/d} dx + C_2 \epsilon^{p/d} + C_3 \epsilon^{(d-p)/d}.$$

As ϵ tends to zero, we see that m tends to infinity and $\cup_{i \in I-J} Q_i \uparrow [0,1]^d$. We apply the monotone convergence theorem to conclude that

(7.11) $$\limsup_{n \to \infty} EL^p(X_1, ..., X_n)/n^{(d-p)/d} \leq \alpha(L_B^p, d) \int \phi(x)^{(d-p)/d} dx.$$

We have now established the upper bound implicit in (7.3). Establishing the lower bound implicit in (7.3) does not involve any new ideas. We merely use the superadditivity of L_B in place of the subadditivity of L. This convenience is not accidental and in fact is one of the motivating reasons for considering boundary functionals in the first place.

Therefore, by the smoothness and the superadditivity of the boundary functional L_B^p, we obtain the lower estimate

$$EL_B^p(X_1, ..., X_n) \geq \sum_{i \in I-J} EL_B^p(\{X_1, ..., X_n\} - A \cap Q_i, Q_i) -$$
$$- C_3(\epsilon n)^{(d-p)/d}.$$

Using this bound, *following the analysis of (7.10) through (7.11) verbatim*, and using Theorem 4.1 once more, we deduce the analogous lower bound

(7.12) $$\liminf_{n \to \infty} EL_B^p(X_1, ..., X_n)/n^{(d-p)/d} \geq \alpha(L_B^p, d) \int_{[0,1]^d} \phi(x)^{(d-p)/d} dx.$$

Combining the complementary estimates (7.11) and (7.12) and using the closeness in mean (3.15) we see that the limsup in (7.11) and the liminf in (7.12) coincide.

Therefore, the asymptotics (7.3) hold for blocked distributions, as desired. By the reduction Lemma 7.3, this concludes the proof of Theorem 7.1.   □

7.3.   Extensions of the Umbrella Theorem

The previous section proved a generalization of the remarkable Beardwood, Halton, and Hammersley (1959) theorem describing the shortest tour on random points. This general umbrella result covers a lot of ground. As we will soon see, it satisfactorily describes the asymptotic behavior of the lengths of graphs given by a wide range of problems in geometric probability. The only shortcoming is that the theorem is limited by the assumption that the underlying point sets have compact support.

In this section we remove this shortcoming and consider graphs on unbounded random point sets in $\mathbb{R}^d$, $d \geq 2$. Our main result, Theorem 7.6, is straightforward and easy to state, but the proof involves a rather lengthy and difficult computation. The reader may skip the proof without loss of continuity.

We draw our inspiration from Rhee (1993a), who determined the asymptotics for the TSP functional over unbounded domains. In the process she disproved a conjecture of Beardwood, Halton, and Hammersley (1959). Rhee's approach depends upon a definition.

Definition 7.4. Let A_o denote the ball in $\mathbb{R}^d$ centered at the origin and with radius 2. For all $k \geq 1$, let A_k denote the annular shell centered around A_o with inner radius 2^k and outer radius 2^{k+1}. Given $f \in L^1(\mathbb{R}^d)$ set

$$a_k(f) := 2^{dk/(d-1)} \int_{A_k} f(x)dx.$$

Rhee's (1993a) main theorem is as follows.

Theorem 7.5. *(asymptotics for the TSP on unbounded domains) Let* $(X_i)_{i \geq 1}$ *be i.i.d. random variables with an absolutely continuous distribution on* $\mathbb{R}^d$ *having a density* $f(x)$. *If* $\int_{\mathbb{R}^d} f(x)^{(d-1)/d}dx < \infty$ *and*

$$(7.13) \qquad\qquad \sum_{k=1}^{\infty}(a_k(f))^{(d-1)/d} < \infty$$

then the TSP functional T *satisfies*

$$(7.14) \qquad \lim_{n \to \infty} T(X_1, ..., X_n)/n^{(d-1)/d} = \alpha(T_B, d) \int_{\mathbb{R}^d} f(x)^{(d-1)/d}dx \quad a.s.$$

Remarks.

(i) Condition (7.13) is satisfied whenever the density f satisfies the moment condition $\int_{\mathbb{R}^d} |x|^r f(x)dx < \infty$ for some $r > d/(d-1)$. To see this, let $\epsilon > 0$ be fixed and consider the representation

$$\sum_{k=1}^{\infty} (a_k(f))^{(d-1)/d} = \sum_{k=1}^{\infty} 2^{-\epsilon k} \left(\int_{A_k} f(x)dx \right)^{(d-1)/d} 2^{(1+\epsilon)k}.$$

Applying Hölder's inequality yields the upper bound

$$\leq C \left(\sum_{k=1}^{\infty} (\int_{A_k} f(x)dx) \cdot 2^{\frac{(1+\epsilon)kd}{d-1}} \right)^{(d-1)/d}$$

$$\leq C \left(\sum_{k=1}^{\infty} \int_{A_k} |x|^{\frac{(1+\epsilon)d}{d-1}} f(x)dx \right)^{(d-1)/d}$$

$$= C \left(\int |x|^{\frac{(1+\epsilon)d}{d-1}} f(x)dx \right)^{(d-1)/d}.$$

Thus (7.13) holds whenever $\int |x|^r f(x)dx < \infty$ for $r > d/(d-1)$.

(ii) The best condition on the sequence $a_k(f)$ that will insure (7.14) is the condition (7.13). In other words, consider a sequence $a_k > 0$ such that $\sum_{k=1}^{\infty} a_k^{(d-1)/d} = \infty$. Then, as shown by Rhee (1993a), there is an f such that $\int_{\mathbb{R}^d} f(x)^{(d-1)/d} dx < \infty$ and $a_k(f) \leq a_k$ for which

$$\lim_{n \to \infty} T(X_1, ..., X_n)/n^{(d-1)/d} = \infty.$$

The condition (7.13) is however not necessary and it is unknown whether there is a meaningful necessary and sufficient condition for (7.14).

By following Rhee's approach in the context of sub and superadditive Euclidean functionals, we will prove a version of Theorem 7.1 which holds for random variables with unbounded support. Since we will work with the general case $1 \leq p < d$ this suggests redefining $a_k(f)$ by

$$a_k(f) := a_{k,p}(f) := 2^{dkp/d-p} \int_{A_k} f(x)dx.$$

Our generalized umbrella theorem takes the following form. By following the ideas of Remark (i) it is easy to check that the theorem applies to i.i.d. random variables having a density f satisfying the integrability condition $\int |x|^r f(x)dx < \infty$ for some $r > d/d - p$.

Notice that if L^p is a Euclidean functional and L^p_B is the canonical boundary Euclidean functional, then $L^p_B(F, \mathbb{R}^d)$ and $L^p(F, \mathbb{R}^d)$ coincide. We will assume that $L^p(F, \mathbb{R}^d) = L^p(F, R)$ whenever $F \cap R = F$. In the remainder of the chapter we write $L^p(F)$ and $L^p_B(F)$ for $L^p(F, \mathbb{R}^d)$ and $L^p_B(F, \mathbb{R}^d)$, respectively.

Theorem 7.6. *(umbrella theorem for Euclidean functionals on $\mathbb{R}^d$, $d \geq 2$) Let L^p and L_B^p be smooth subadditive and superadditive Euclidean functionals of order p, respectively. Assume that L^p satisfies simple subadditivity (2.2) and that L^p and L_B^p are close in mean (3.15). Let $(X_i)_{i \geq 1}$ be i.i.d. random variables with an absolutely continuous distribution on $\mathbb{R}^d$ having a density $f(x)$. If $1 \leq p < d$, $\int_{\mathbb{R}^d} f(x)^{(d-p)/d} dx < \infty$, and*

$$(7.15) \qquad \sum_{k=1}^{\infty} (a_k(f))^{(d-p)/d} < \infty,$$

then the Euclidean functional L^p satisfies

$$(7.16) \qquad \lim_{n \to \infty} L^p(X_1, ..., X_n)/n^{(d-p)/d} = \alpha(L_B^p, d) \int_{\mathbb{R}^d} f(x)^{(d-p)/d} dx \quad a.s.$$

It is easy to see that the TSP functional T satisfies the conditions of Theorem 7.6 and thus Theorem 7.6 generalizes Theorem 7.5. Moreover, it is also easy to check that both the minimal spanning tree functional M and the minimal matching functional S satisfy the hypotheses of Theorem 7.6 and therefore (7.16) provides the asymptotics for the solutions to the archetypical problems of combinatorial optimization over samples of unbounded support. Later we will see that other problems in geometric probability satisfy the hypotheses of Theorem 7.6. This includes the k-median problem and the semi-matching problem among others.

Let us now turn to the proof of the umbrella Theorem 7.6. Throughout we will closely follow the proof of Rhee's Theorem 7.5. There are few new ideas and the proof is unfortunately rather long and technical. We have included it for completeness.

Lower bounds for Euclidean functionals are often the most difficult, but now to prove the lower bound implicit in (7.16), namely the bound

$$(7.17) \qquad \liminf_{n \to \infty} L_B^p(X_1, ..., X_n)/n^{(d-p)/d} \geq \alpha(L_B^p, d) \int_{\mathbb{R}^d} f(x)^{(d-p)/d} dx \quad a.s.$$

we have only to make a few elementary observations. Given $\epsilon > 0$, let $Q := Q(\epsilon)$ be a cube centered at the origin such that $\rho := P(Q) \geq 1 - \epsilon$, where P is the law of X. Using superadditivity as a surrogate for monotonicity we have

$$\liminf_{n \to \infty} L_B^p(X_1, ..., X_n)/n^{(d-p)/d} \geq \liminf_{n \to \infty} L_B^p(\{X_1, ..., X_n\} \cap Q, Q)/n^{(d-p)/d}.$$

Now let Z denote the restriction of X to the cube Q. Then by the smoothness of L_B^p, the above is a.s. bounded below by

$$(7.18) \qquad \liminf_{n \to \infty} L_B^p(Z_1, ..., Z_{n\rho}, Q)/n^{(d-p)/d}.$$

Since the random variables Z have compact support, Theorem 7.1 implies that (7.18) is in turn bounded below by

$$(7.19) \qquad \rho^{(d-p)/d} \alpha(L_B^p, d) \int_{\mathbb{R}^d} f_Z(x)^{(d-p)/d} dx \quad a.s.,$$

where f_Z denotes the density for the random variable Z. Let ϵ go to zero in (7.19) and apply Fatou's lemma to deduce the lower bound (7.17).

The proof of the lower bound (7.17) is rather straightforward. The proof of (7.16) thus depends on the proof of the upper bound implicit in (7.16), i.e., the proof of

$$(7.20) \qquad \limsup_{n \to \infty} L^p(X_1, ..., X_n)/n^{(d-p)/d} \leq \alpha(L_B^p, d) \int_{\mathbb{R}^d} f(x)^{(d-p)/d} dx \quad a.s.$$

The proof of (7.20) is broken into two steps and we will faithfully follow Rhee (1993a).

We set the stage by letting $s(n)$ be the largest $k \in \mathbb{N}$ such that $A_k \cap \{X_1, ..., X_n\}$ is not empty; for all $1 \leq q \leq s(n)$ we let B_q be the ball of radius 2^q. We now bound $L^p(X_1, ..., X_n)$ by considering the simple subadditivity (2.2) of the Euclidean functional L^p over the union of $A_{s(n)}$ and $\{\bigcup_{k=q}^{s(n)-1} A_k \cup B_q\}$. Simple subadditivity gives for fixed q

$$L^p(X_1, ..., X_n) \leq L^p\left(\{X_1, ..., X_n\} \cap A_{s(n)}\right) +$$
$$+ L^p\left(\{X_1, ..., X_n\} \cap \{\cup_{k=q}^{s(n)-1} A_k \cup B_q\}\right) + C_1(2^{s(n)+1})^p.$$

Applying simple subadditivity to the second term on the right side of the above gives

$$L^p\left(\{X_1, ..., X_n\} \cap \{\bigcup_{k=q}^{s(n)-1} A_k \cup B_q\}\right)$$
$$\leq L^p\left(\{X_1, ..., X_n\} \cap A_{s(n)-1}\right) +$$
$$+ L^p\left(\{X_1, ..., X_n\} \cap \{\bigcup_{k=q}^{s(n)-2} A_k \cup B_q\}\right) + C_1(2^{s(n)})^p.$$

Repeatedly applying simple subadditivity and bounding the geometric series $\sum_{k=q}^{s(n)} 2^{kp}$ by $C(p)2^{ps(n)}$ we come to the starting point of the proof of (7.20), namely we arrive at

$$(7.21) \qquad L^p(X_1, ..., X_n) \leq L^p(\{X_1, ..., X_n\} \cap B_q) +$$
$$+ \sum_{q \leq k \leq s(n)} L^p(\{X_1, ..., X_n\} \cap A_k) + C(p)2^{ps(n)}.$$

The proof of (7.20) now follows from the following two steps. Since $L^p(F, \mathbb{R}^d) = L^p(F, R)$ if $F \cap R = F$, growth bounds for Euclidean functionals imply that if

$$N_k(n) := \text{card}(\{X_1, ..., X_n\} \cap A_k),$$

then $L^p(\{X_1, ..., X_n\} \cap A_k) \leq C 2^{kp}(N_k(n))^{(d-p)/d}$.

Step 1.   Show that

$$(7.22) \qquad \limsup_{n \to \infty} \frac{L^p(\{X_1, ..., X_n\} \cap B_q)}{n^{(d-p)/d}} \le \alpha(L_B^p, d) \int_{B_q} f(x)^{(d-p)/d} dx \quad a.s.$$

Step 2.   Show that

$$(7.23) \quad \limsup_{n \to \infty} \left\{ C(p) 2^{p \cdot s(n)} + \sum_{k=q}^{s(n)} 2^{kp} (N_k(n))^{(d-p)/d} \right\} / n^{(d-p)/d} \le t(p, q) \quad a.s.,$$

where $\lim_{q \to \infty} t(p, q) = 0$.

Indeed, combining (7.21)-(7.23) and letting q tend to infinity yields the desired upper bound (7.20).

Let us now prove (7.22). Fix q and set

$$\int_{B_q} f(x) dx = 1 - \epsilon$$

where ϵ is between 0 and 1. We let Z denote the restriction of the random variable X to the ball B_q, we let f_Z be its density, and we let $B(n, 1 - \epsilon)$ denote a binomial random variable with parameters n and $1 - \epsilon$. By the assumed smoothness of L^p we have with probability one the estimate

$$\limsup_{n \to \infty} \frac{L^p(\{X_1, ..., X_n\} \cap B_q)}{n^{(d-p)/d}}$$

$$= \limsup_{n \to \infty} \frac{L^p(Z_1, ..., Z_{B(n,1-\epsilon)})}{n^{(d-p)/d}}$$

$$\le \limsup_{n \to \infty} \frac{L^p(Z_1, ..., Z_{n(1-\epsilon)})}{n^{(d-p)/d}} +$$

$$+ \, C_3 2^{qp} \limsup_{n \to \infty} \frac{|B(n, 1 - \epsilon) - n(1 - \epsilon)|^{(d-p)/d}}{n^{(d-p)/d}}$$

$$= (1 - \epsilon)^{(d-p)/d} \alpha(L_B^p, d) \int_{B_q} f_Z(x)^{(d-p)/d} dx,$$

where the last equality follows from the strong law of large numbers. The proof of (7.22) and Step 1 is completed by noting that $f_Z(x) = f(x)/(1 - \epsilon)$ for all $x \in B_q$.

To prove Step 2 we only need to bound the tail of binomial distributions. This involves some lengthy computations and we will follow the approach used by Rhee (1993a) for the TSP and generalized by McGivney (1997).

To prove (7.23) we note that since $N_k(n)$ and $s(n)$ increase in n, it suffices to show with probability one that

$$(7.24) \qquad \limsup_{r \to \infty} T(r,p,q) \le C(p) \sum_{k \ge q} 2^{-p \cdot t(k)},$$

where

$$T(r,p,q) := \frac{1}{2^{r(d-p)/d}} \left\{ C(p) 2^{p \cdot s(2^r)} + \sum_{k \ge q} 2^{kp} (N_k(2^r))^{(d-p)/d} \right\},$$

and where $\sum_{k \ge 1} 2^{-p \cdot t(k)}$ is a convergent sum.

To prove (7.24) we fix p, $1 \le p < d$, and for $r \in \mathbb{N}$ define events E_r in such a way that on the infinite intersection $\bigcap_{r \ge m} E_r$, m arbitrary, we have

$$(7.25) \qquad \limsup_{r \to \infty} T(r,p,q) \le C \sum_{k \ge q} 2^{-p \cdot t(k)}$$

and moreover

$$(7.26) \qquad \lim_{m \to \infty} P(\bigcap_{r \ge m} E_r) = 1.$$

We will choose $E_r := \bigcap_{i=1}^{3} E_{i,r}$, where the events $E_{1,r}, E_{2,r}$, and $E_{3,r}$ are defined as follows and where the integers A, B, Q, and R are to be chosen later, all as a function of the parameter r:

$$E_{1,r} := \{\forall k > B, N_k(2^r) = 0\}$$

and

$$E_{2,r} := \{ \sum_{A \le k \le B} N_k(2^r) \le Q\}$$

and

$$E_{3,r} := \{\forall k < A, N_k(2^r) \le R\}.$$

Let us first consider how to arrange for (7.26). Set $m_k := \int_{A_k} f(x)dx$ so that

$$a_k(f) := 2^{kdp/(d-p)} m_k.$$

Clearly, if for each $1 \le i \le 3$ we have $\sum_{r=1}^{\infty} P\{E_{i,r}^c\} < \infty$ then the limit (7.26) follows. We will now show that the sums $\sum_{r=1}^{\infty} P\{E_{i,r}^c\}, 1 \le i \le 3$, are each bounded by an expression defined in terms of the $m_k, k \ge 1$, which we will in turn bound by terms from the sum $\sum_{k=1}^{\infty} 2^{-t(k)}$. We have by independence

$$P\{E_{1,r}\} = P\left\{ \bigcap_{k>B} \{X_1 \notin A_k, ..., X_{2^r} \notin A_k\} \right\}$$

$$= \left(P\{ \bigcap_{k>B} (X_1 \notin A_k) \} \right)^{2^r}$$

$$(7.27) \qquad = (1 - \sum_{k>B} m_k)^{2^r}.$$

Considering $P\{E_{2,r}^c\}$, we have

$$P\{E_{2,r}^c\} = P\{\sum_{A \leq k \leq B} N_k(2^r) > Q\} \leq P\{B(2^r, \sum_{A \leq k \leq B} m_k) \geq Q\}.$$

By standard estimates for the tail of a binomial random variable (see e.g. Shorack and Wellner (1986,Chapter 11)), the right side of the above is bounded by $\exp(-\alpha Q)$ if $Q \geq 2^{r+1} \sum_{A \leq k \leq B} m_k$. Here α is a generic positive constant whose value is not important to us. Thus

(7.28) $$P\{E_{2,r}^c\} \leq \exp(-\alpha Q)$$

if Q is at least as large as $2^{r+1} \sum_{A \leq k \leq B} m_k$. Considering $P\{E_{3,r}^c\}$ we have

(7.29) $$P\{E_{3,r}^c\} \leq \sum_{k < A} P\{B(2^r, m_k) > R\} \leq \sum_{k < A} \exp(-\alpha R)$$

if R is at least as large as $2^{r+1} m_k$. Thus the three probabilities $P\{E_{i,r}^c\}, 1 \leq i \leq 3$, are each bounded by terms involving m_k.

We are thus motivated to search for bounds for the terms m_k, $k \geq 1$. Since $m_k := a_k(f) 2^{-kdp/(d-p)}$ and since the behavior of the terms $a_k(f)$ may be irregular we will regularize the sequence $a_k(f)$, $k \geq 1$, and find a sequence $t(k)$, $k \geq 1$, of positive numbers which satisfies

(7.30) $$a_k(f)^{(d-p)/d} \leq 2^{-p \cdot t(k)},$$

(7.31) $$\sum_{k=1}^{\infty} 2^{-p \cdot t(k)} < \infty,$$

and

(7.32) $$\forall j, k \in \mathbb{N} \quad |t(j) - t(k)| \leq \frac{p|j - k|}{2d}.$$

One choice of sequence $t(k)$, $k \geq 1$, is that defined by

$$2^{-p \cdot t(k)} := \sum_{l=1}^{\infty} (a_l(f))^{(d-p)/d} 2^{-p^2 |l-k|/2d}.$$

It is straightforward to verify that such a sequence satisfies (7.30)-(7.32); the details are provided at the end of this chapter.

By (7.30), we have $m_k \leq 2^{-(k+t(k))dp/(d-p)}$ which suggests the natural definition

(7.33) $$m_k' := 2^{-(k+t(k))dp/(d-p)}.$$

This immediately yields

(7.34) $$m_{k+1} \leq m_{k+1}' \leq 2^{-p} m_k',$$

where the second inequality uses (7.32). Moreover, we have

$$(7.35) \qquad \sum_{l=k}^{\infty} m_l \le \sum_{l=k}^{\infty} m'_l \le m'_k \sum_{l \ge 1} 2^{-pl} \le 2m'_k.$$

Since $m_k \le m'_k := 2^{-(k+t(k))dp/d-p}$, our search for bounds for m_k brings us to look for bounds for $k + t(k)$. For each $r \in \mathbb{N}$, denote by $k(r)$ the largest k such that

$$(7.36) \qquad k + t(k) \le r\frac{d-p}{dp}.$$

By (7.32) and (7.36) we have the estimate

$$k(r) + t(k(r)) \le r\frac{d-p}{dp} \le k(r) + 1 + t(k(r) + 1)$$

$$(7.37) \qquad\qquad\qquad \le k(r) + t(k(r)) + 2,$$

which implies that if $k(r) = k(r')$ then $\frac{d-p}{dp}|r - r'| \le 2$. Using this observation and (7.31) it is easy to see that we must have

$$(7.38) \qquad \sum_{r=1}^{\infty} 2^{-p \cdot t(k(r))} < \infty.$$

This convergent sum will serve us well in our ongoing attempt to bound the three sums $\sum_{r \ge 1} P\{E^c_{i,r}\}$, $i = 1, 2, 3$. More precisely, we will show for each $i = 1, 2, 3$ that

$$P\{E^c_{i,r}\} \le C2^{-p \cdot t(k(r))}.$$

Let's first consider the case $i = 1$. From (7.27) it is clearly desirable to obtain a bound of the form

$$(7.39) \qquad \sum_{k>B} m_k \le 2^{-p \cdot t(k(r))-r+C},$$

for some appropriate choice of C. Now (7.35) gives

$$(7.40) \qquad \sum_{k>B} m_k \le 2m'_{B+1} = 2 \cdot 2^{-(B+1+t(B+1))\frac{dp}{d-p}}.$$

Notice that if B is defined by

$$(7.41) \qquad B := [k(r) + (1 - b)t(k(r))],$$

where $0 < b < 1$ is a constant to be chosen later, then we can achieve the necessary small upper bound (7.39). Noting that (7.32) yields

$$|t(B + 1) - t(k(r) + 1)| \le \frac{p(B - k(r))}{2d} \le \frac{p \cdot t(k(r))}{2d}$$

we indeed find by (7.40) and (7.41) that

$$\sum_{k>B} m_k \le 2 \cdot 2^{-\left(k(r)+(1-b)t(k(r))+t(k(r)+1)-\frac{p(B-k(r))}{2d}\right)\frac{dp}{d-p}}$$

$$\le 2 \cdot 2^{-\left(k(r)+(1-b)t(k(r))+t(k(r)+1)-\frac{p \cdot t(k(r))}{2d}\right)\frac{dp}{d-p}}$$

$$\le 2 \cdot 2^{-\left(r(\frac{d-p}{dp})-1+(1-b-\frac{p}{2d})t(k(r))\right)\frac{dp}{d-p}},$$

where the last inequality follows by (7.37). Many values of b will achieve the bound (7.39); for specificity we choose $b := p/2d$ and obtain the rough estimate

$$\sum_{k>B} m_k \le 2 \cdot 2^{-r+\frac{dp}{d-p}-p \cdot t(k(r))}.$$

Returning to (7.27) and using the estimate $e^{-x} \ge 1 - x$ for x small and positive gives

$$P\{E_{1,r}\} \ge 1 - C \cdot 2^{\frac{dp}{d-p}-p \cdot t(k(r))}$$

and therefore

$$P\{E_{1,r}^c\} \le C \cdot 2^{-p \cdot t(k(r))},$$

as desired.

Now we turn to an estimate for $P\{E_{2,r}^c\}$. We recall by (7.28) that

$$P\{E_{2,r}^c\} \le \exp(-\alpha Q)$$

if $Q := Q(r) \ge 2^{r+1} \sum_{A \le k \le B} m_k$. Observe by (7.35) that

$$(7.42) \qquad 2^{r+1} \sum_{A \le k \le B} m_k \le 2^{r+2} 2^{-(A+t(A))\frac{dp}{d-p}}.$$

We would like to select A in such a way that we can choose $Q(r) \approx 2^{Ct(k(r))}$ and thus obtain

$$(7.43) \qquad P\{E_{2,r}^c\} \le \exp(-\alpha Q) \le C \cdot 2^{-p \cdot t(k(r))}.$$

As with the choice of B, we see that if A is of the form

$$(7.44) \qquad A := [k(r) - at(k(r))],$$

where $0 < a < 1$ is a constant to be chosen, then we can apply (7.37) and achieve the bound (7.42). Noting that (7.32) yields

$$|t(A) - t(k(r))| \le \frac{p(k(r) - A)}{2d} \le \frac{p \cdot t(k(r))}{2d},$$

we indeed find by (7.42) and (7.44) that

$$2^{r+1} \sum_{A \le k \le B} m_k \le 2^{r+2} 2^{-\left(k(r)-at(k(r))+t(k(r))-\frac{p(k(r)-A)}{2d}\right)\frac{dp}{d-p}}.$$

By (7.37) we obtain $|k(r) + t(k(r)) - r\frac{d-p}{dp}| \leq 2$ and thus

$$2^{r+1} \sum_{A \leq k \leq B} m_k \leq C2^r 2^{-(r\frac{d-p}{dp} - (a+\frac{ap}{2d})t(k(r)))\frac{dp}{d-p}}$$

$$\leq C2^{2at(k(r))\frac{dp}{d-p}},$$

since $a + \frac{ap}{2d} \leq 2a$. We now see that if we set

$$Q := C2^{2at(k(r))\frac{dp}{d-p}},$$

then (7.28) gives the desired estimate

$$\sum_r P\{E_{2,r}^c\} \leq \sum_r \exp(-\alpha Q) \leq C \sum_r 2^{-p \cdot t(k(r))} < \infty,$$

since $\exp(-\alpha 2^{-ct}) \leq 2^{-pt}$ for t large enough.

To conclude the proof of (7.26) it only remains to verify that $\sum_r P\{E_{3,r}^c\} < \infty$. We will follow the ideas above. We recall by (7.29) that

$$P\{E_{3,r}^c\} \leq \sum_{k < A} \exp(-\alpha R)$$

if $R := R(r) \geq 2^{r+1} m_k \geq 2^{r+1} m'_k$. We may thus choose $R := 2^{r+1} m'_k$ and we now seek lower bounds on R in order to bound $\sum_{k<A} \exp(-\alpha R)$. Observe that iteration of the estimate $m'_k \geq 2^p m_{k+1}$ a total of $A - k$ times gives

$$m'_k \geq 2^{p(A-k)} m'_A.$$

Thus by (7.33) and (7.44) we have for $k < A$

$$m'_k \geq 2^{p(A-k)} 2^{-(A+t(A))\frac{dp}{d-p}}$$

$$\geq 2^{p(A-k)} 2^{-(k(r)-at(k(r))+1+t(k(r))+\frac{p(A-k(r))}{2d})\frac{dp}{d-p}}.$$

Arguing as in previous cases, we obtain from (7.37)

$$m'_k \geq 2^{p(A-k)} 2^{-(k(r)-at(k(r))+1+t(k(r))+\frac{p(1-at(k(r)))}{2d})\frac{dp}{d-p}}$$

$$\geq 2^{p(A-k)} 2^{-(r\frac{d-p}{dp}+2-t(k(r))(a+\frac{ap}{2d}))\frac{dp}{d-p}}$$

$$\geq C2^{p(A-k)} 2^{-r} 2^{(a+\frac{ap}{2d})\frac{dp}{d-p}}$$

$$\geq C2^{p(A-k)} 2^{-r+ap \cdot t(k(r))},$$

where we use $-p/2d > -1$, $a + ap/2d > a$, and $\frac{dp}{d-p} \geq p$. Thus

$$R \geq C2^{p(A-k)+ap \cdot t(k(r))}.$$

It follows that

$$P\{E^c_{3,r}\} \le \sum_{k<A} \exp(-\alpha R)$$

$$\le \sum_{k<A} \exp(-\alpha C 2^{p(A-k)} 2^{ap \cdot t(k(r))})$$

$$\le C \exp(-\alpha C 2^{ap \cdot t(k(r))}).$$

Now since $\sum_r 2^{-p \cdot t(k(r))} < \infty$ we have $\lim_{r \to \infty} t(k(r)) = \infty$ and thus if r is large enough

$$\exp(-\alpha C 2^{ap \cdot t(k(r))}) \le 2^{-p \cdot t(k(r))}$$

which gives

$$\sum_r P\{E^c_{3,r}\} \le C \sum_r 2^{-p \cdot t(k(r))} < \infty.$$

We have thus shown the equality (7.26) for $A := A(r) := [k(r) - at(k(r))]$, $B := [k(r) + (1 - \frac{p}{2d})t(k(r))]$, $Q := C 2^{2at(k(r))\frac{dp}{d-p}}$, and $R := 2^{r+1}m'_k$, where $a > 0$ is still to be defined. We now show the inequality (7.25).

To see that (7.25) holds, we will consider the behavior of $T(r, p, q)$ on the events $E_{i,r}$, $i = 1, 2, 3$. On $E_{3,r}$ we have by choice of R that

$$2^{-r\frac{d-p}{d}} \sum_{q \le k < A} 2^{kp}(N_k(2^r))^{(d-p)/d}$$

$$\le 2^{-r\frac{d-p}{d}} \sum_{q \le k < A} 2^{kp}(2^{r+1}m'_k)^{(d-p)/d}$$

$$= 2^{-r\frac{d-p}{d}} \sum_{q \le k < A} 2^{kp} \left(2^{r+1} 2^{-(k+t(k))\frac{dp}{d-p}}\right)^{(d-p)/d}$$

(7.45) $$\le C \sum_{q \le k < A} 2^{-p \cdot t(k)}.$$

On the event $E_{2,r}$ we have by Hölder's inequality that

$$2^{-r\frac{d-p}{d}} \sum_{A \le k \le B} 2^{kp}(N_k(2^r))^{(d-p)/d}$$

$$\le 2^{-r\frac{d-p}{d}} \left(\sum_{A \le k \le B} 2^{dk}\right)^{p/d} \left(\sum_{A \le k \le B} N_k(2^r)\right)^{(d-p)/d}$$

$$\le C 2^{-r\frac{d-p}{d}} 2^{Bp} 2^{2ap \cdot t(k(r))},$$

by the definition of Q. Recalling the definition of B, the above is bounded by

$$\le C 2^{-r(\frac{d-p}{d})} 2^{p(k(r) + (1 - \frac{p}{2d})t(k(r))) + 2ap \cdot t(k(r))}$$

$$\le C 2^{-r(\frac{d-p}{d})} 2^{p(r\frac{d-p}{dp} - (\frac{p}{2d} - 2a)t(k(r)))}$$

$$= C 2^{-p(\frac{p}{2d} - 2a)t(k(r))}.$$

Now we choose $a > 0$ such that $\frac{p}{2d} - 2a > 0$. There is more than one value for a which is suitable and for specificity we choose $a := \frac{p}{5d}$, giving $\frac{p}{2d} - 2a = \frac{p}{10d}$. On $E_{2,r}$ this yields the upper bound

$$(7.46) \qquad 2^{-r\frac{d-p}{d}} \sum_{A \leq k \leq B} 2^{kp} \left(N_k(2^r)\right)^{(d-p)/d} \leq C 2^{-\frac{p^2}{10d} t(k(r))}.$$

We note for future reference that the right side of (7.46) approaches zero as r tends to infinity.

Finally, on the event $E_{1,r}$ we have

$$(7.47) \qquad 2^{-r\frac{d-p}{d}} \sum_{k > B} 2^{kp} (N_k(2^r))^{(d-p)/d} = 0.$$

To estimate the remaining term $2^{p \cdot s(2^r)} 2^{-r\frac{d-p}{d}}$ in $T(r,p,q)$ we note that on $E_{1,r}$ we have $N_k(2^r) = 0$ for all $k > B$, implying $s(2^r) \leq B$. Thus by the definition of B we obtain the upper bound

$$2^{p \cdot s(2^r) - r\frac{d-p}{d}} \leq 2^{pB - r\frac{d-p}{d}}$$
$$\leq 2^{p[k(r) + (1 - \frac{p}{2d})t(k(r))] - r\frac{d-p}{d}}$$
$$= 2^{-\frac{p^2 t(k(r))}{2d}},$$

from which it follows that

$$(7.48) \qquad \lim_{r \to \infty} 2^{p \cdot s(2^r) - \frac{r(d-p)}{d}} = 0.$$

Collecting the estimates (7.45) - (7.48) we obtain on the set $\bigcap_{r \geq m} E_r$, $m \in \mathbb{N}$ and arbitrary, the upper bound

$$\limsup_{r \to \infty} 2^{-\frac{r(d-p)}{d}} \left\{ C 2^{p \cdot s(2^r)} + \sum_{k \geq q} 2^k (N_k(2^r))^{(d-p)/d} \right\} \leq C \sum_{q \leq k < A} 2^{-p \cdot t(k)}.$$

This is precisely (7.24), which was to be shown. This completes the proof of Theorem 7.6. $\square$

Notes and References

1. The proof of the umbrella Theorem 7.1 follows Redmond and Yukich (1994) which treats the case $p = 1$ and Redmond and Yukich (1996), which treats the general case $1 \leq p < d$. The proof of the umbrella Theorem 7.6 follows approximately as in McGivney (1997). It would be worthwhile to find an umbrella theorem for values of p in the range $0 < p < 1$ and $p \geq d$.

2. Smoothness of L is not necessary in order to insure (7.2). Indeed, the bipartite matching functional is not smooth but nonetheless satisfies (7.2); see Dobrić and Yukich (1995). See Yukich (1992) for asymptotics for bi-partite matching on $\mathbb{R}^d$.

3. We verify that the sequence $t(k)$, $k \geq 1$, satisfies conditions (7.30)-(7.32). To see that condition (7.30) is satisfied note simply that

$$2^{-t(k)} = \sum_{l=1}^{\infty} a_l(f)^{(d-p)/dp} \, 2^{-p|l-k|/2d}$$

$$\geq a_k(f)^{(d-p)/dp}.$$

To verify (7.31) we observe that

$$\sum_{k=1}^{\infty} 2^{-p \cdot t(k)} = \sum_{k=1}^{\infty} \sum_{l=1}^{\infty} (a_l(f))^{(d-p)/d} 2^{-p^2|l-k|/2d}$$

$$< \sum_{k=-\infty}^{\infty} \sum_{l=1}^{\infty} (a_l(f))^{(d-p)/d} 2^{-p^2|l-k|/2d}$$

$$= \sum_{l=1}^{\infty} (a_l(f))^{(d-p)/d} \sum_{k=-\infty}^{\infty} 2^{-p^2|l-k|/2d}$$

$$< \infty.$$

Finally, to show (7.32) it suffices to show both

$$t(k) - t(j) \leq \frac{p|j-k|}{2d} \quad \text{and} \quad t(j) - t(k) \leq \frac{p|j-k|}{2d}.$$

We will show the first inequality; the second holds in a similar way. To show the first inequality, it suffices to show

$$2^{-t(j)-\frac{p|j-k|}{2d}} \leq 2^{-t(k)}.$$

Observe that this follows from

$$2^{-t(k)} = \sum_{l=1}^{\infty} (a_l(f))^{(d-p)/dp} 2^{-\frac{p|l-k|}{2d}}$$

$$\geq \sum_{l=1}^{\infty} (a_l(f))^{(d-p)/dp} 2^{-\frac{p}{2d}(|l-j|+|j-k|)}$$

$$= 2^{-\frac{p}{2d}|j-k|} \sum_{l=1}^{\infty} (a_l(f))^{(d-p)/dp} 2^{-\frac{p}{2d}|l-j|}$$

$$= 2^{-\frac{p}{2d}|j-k|} 2^{-t(j)}.$$

The second inequality follows similarly.

8. APPLICATIONS AND EXAMPLES

We have seen that the complementary notions of subadditivity and superadditivity, together with isoperimetric methods, are well suited for proving limit theorems for the lengths of Euclidean graphs. We used these tools, together with a smoothness regularity condition, to obtain laws of large numbers, rates of convergence, and large deviations for the total edge lengths of graphs.

The previous chapters showed that our general methods describe the asymptotics of the lengths of the graphs of several of the outstanding problems of geometric probability. This includes the archetypical problems of combinatorial optimization, namely the length of the shortest tour on a random sample, the minimal length of a tree spanned by a random sample, and the length of a minimal Euclidean matching on a random sample.

The remainder of this monograph furnishes additional examples of graphs whose lengths may be studied by our general methods. We show that the general approach provides limit theorems for a wide variety of problems in combinatorial optimization, including geometric location problems on a random sample, the many traveling salesman problem on a random sample, and the semi-matching problem on a random sample.

We show that the general approach treats other classes of problems, especially the lengths of the graphs of some of the fundamental problems in computational geometry. This includes the length of the k nearest neighbors graph on a random sample. We will also show that the approach furnishes asymptotics for graphs occurring in minimal surfaces, particularly the length of the minimal triangulation of a random sample. Our examples are intended for purposes of illustration and they are not supposed to be exhaustive.

8.1. Steiner Minimal Spanning Trees

Let $F := \{x_1, x_2, ..., x_n\} \subset \mathbb{R}^d$, $d \geq 2$. A Steiner tree on F is a connected graph which contains F. The graph may include "Steiner points", that is vertices other than those in F. For example if F consists of the vertices of an equilateral triangle then the Steiner minimal spanning tree is obtained by joining all three vertices to the Steiner point lying in the center of the triangle. See Figure 8.1.

It is well-known that Steiner minimal spanning trees exist. This is a consequence of the easily proved fact that all vertices in Euclidean minimal spanning trees have bounded degree (see e.g. Melzak (1973)). More generally, the length of a Steiner minimal spanning tree on F with pth power-weighted edges is defined by

$$\hat{M}^p(F) := \min_S \sum_{e \in S} |e|^p,$$

where the minimum ranges over all Steiner trees S on F.

Figure 8.1. A Steiner minimal spanning tree (three vertices connected to a central Steiner point)

The Steiner MST functional is a smooth subadditive Euclidean functional. Indeed, if $\hat{M}^p(F, R)$ denotes the Steiner MST functional on pairs $(F, R) \in \mathcal{F} \times \mathcal{R}$, then for $F_1 \subset F_2$ we have $\hat{M}^p(F_1, R) \leq \hat{M}^p(F_2, R)$, which shows monotonicity. Using monotonicity and the simple subadditivity of $\hat{M}^p$ it is easy to verify that $\hat{M}^p$, $0 < p < d$, is a smooth subadditive Euclidean functional of order p.

The associated boundary Steiner MST functional $\hat{M}^p_B$, $1 \leq p < d$, is defined in the natural way and it is likewise easy to check that $\hat{M}^p_B$ is a superadditive Euclidean functional of order p. Checking pointwise closeness, closeness of means, and smoothness may be done exactly as in Chapter 3. The Steiner MST functional thus satisfies the conditions of the basic limit Theorem 4.1 as well as the umbrella Theorem 7.1.

8.2. Semi-Matchings

A "tour" or "cycle" is a connected graph in which all points have degree two. If we drop the requirement of connectedness then we have what is sometimes called a 2-matching or semi-matching. A minimal 2-matching on the vertex set $V :=$ $\{v_1, ..., v_n\}$ thus produces a graph on V of minimal total edge length in which all vertices have degree 2 with the understanding that an isolated edge between two vertices v_1 and v_2 actually represents two copies of the edge $v_1 v_2$. The graph thus contains cycles with an odd number of edges ("odd cycles") as well as isolated edges. See Figure 8.2.

Figure 8.2. Semi-matchings contain isolated edges and odd cycles

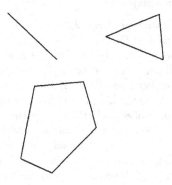

Semi-matchings thus represent a relaxation of the TSP and can be described in the language of linear programming. Let $G = (V, E)$ be a graph such that for each edge $e \in E$ there is an associated weight w_e. We seek solutions to the following linear programming problem:

$$(8.1) \qquad\qquad z = \min_x \sum_{e \in E} x_e w_e$$

subject to the constraints

$$\sum_{e \text{ meets } v} x_e = 1 \text{ for all } v \in V \text{ and } x_e \geq 0 \text{ for all } e \in E.$$

The Euclidean semi-matching problem seeks the solutions to (8.1) where the vertices V of G are points $v_1, ..., v_n$ in $\mathbb{R}^d$ and where the weight w_e associated with edge $e := v_i v_j$ is the Euclidean distance $\|v_i - v_j\|$. Balinski (1965) showed that the loading factor x_e can only be $0, 1/2,$ or 1 in a minimal solution to (8.1). It follows that any minimal solution consists of a union of isolated edges with loading 1 and a collection of odd cycles that has all edge loading equal to $1/2$. See Figure 8.2.

We let $\hat{S}(F)$ denote the Euclidean semi-matching functional on the vertex set F; thus $\hat{S}$ equals the sum of the combined edge lengths of the minimal solution. Notice that the semi-matching functional is related to the TSP functional by $\hat{S}(F) \leq T(F)/2$. It is easy to verify that $\hat{S}$ is a subadditive functional (3.4) and in fact has no error term (that is C_1 is zero in (3.4)).

We now define the canonical boundary functional $\hat{S}_B$ associated with the semi-matching functional $\hat{S}$. This is done in a way which resembles the construction of the canonical boundary matching functional S_B. We define the boundary semi-matching functional by

$$\hat{S}_B(F, R) := \min \left(\hat{S}(F, R), \ \inf \sum_i \hat{S}(F_i \cup \{a_i, b_i\}) \right),$$

where the infimum ranges over all partitions $(F_i)_{i \geq 1}$ of F and all sequences of pairs of points $\{a_i, b_i\}$ lying on the boundary of $[0,1]^d$. The boundary is treated as a single vertex and edges belonging to the boundary are thus not counted when forming the odd cycles of the optimal semi-matching.

The boundary functional is superadditive (3.3). Indeed, the global semi-matching on points F in a rectangle $R := R_1 \cup R_2$ generates a graph in R_1 whose length exceeds that of the semi-matching on $F \cap R_1$. To see this, we modify those components of the global semi-matching which meet the boundary between R_1 and R_2. These components are odd cycles or isolated edges. If they are the latter, we will leave them alone. If an odd cycle meets the boundary, then the restriction to R_1 is either an odd or even cycle (the number of edges in the cycle is the number of non-boundary edges). If it is an odd cycle then we leave it alone. However, if the restriction is an even cycle, then we may delete every other edge in the cycle to obtain a set of isolated edges at no extra cost. In either event, the global semi-matching generates a feasible graph in R_1 which is at least as large as the optimal semi-matching of points in $F \cap R_1$. The same is of course true for rectangle R_2 and superadditivity follows.

To see that $\hat{S}$ is smooth of order 1 (3.8) we proceed as follows. Since $\hat{S}$ is subadditive we evidently have for all subsets F in $[0,1]^d$ the growth bound $\hat{S}(F) \leq C(\mathrm{card}F)^{(d-1)/d}$ by Lemma 3.3. To show smoothness, first observe that for all sets F and G in $[0,1]^d$

$$\hat{S}(F \cup G) \leq \hat{S}(F) + \hat{S}(G) \leq \hat{S}(F) + C(\mathrm{card}G)^{(d-1)/d}.$$

For the reverse inequality, let F_1 denote the set of points of F which are connected to points of G by the semi-matching which realizes $\hat{S}(F \cup G)$ and let $F_2 := F - F_1$. We have $\mathrm{card}F_1 \leq 2\mathrm{card}G$ and by considering the cycles in the optimal semi-matching of $F \cup G$ we can construct a feasible semi-matching of length $S'(F_2)$ on F_2 whose length is bounded by the length of the global optimal semi-matching:

$$\hat{S}(F) \leq S'(F_2) + \hat{S}(F_1) \leq \hat{S}(F \cup G) + C(\mathrm{card}F_1)^{(d-1)/d}.$$

On combining the above inequalities we see that $\hat{S}$ is smooth of order 1:

$$|\hat{S}(F \cup G) - \hat{S}(F)| \leq C(\mathrm{card}G)^{(d-1)/d}.$$

Similar arguments show that $\hat{S}_B$ is also smooth of order 1.

To see that $\hat{S}$ is pointwise close to the boundary functional $\hat{S}_B$ we need only follow the proof of Lemma 3.7 for the simple case $p = 1$. To see that $\hat{S}$ is close in mean (3.15) to $\hat{S}_B$, follow the proof of Lemma 3.10.

We thus see that the semi-matching functional $\hat{S}$ and the boundary functional $\hat{S}_B$ are respectively subadditive and superadditive Euclidean functionals of order 1 and therefore satisfy the asymptotics of the umbrella Theorem 7.1.

8.3. k Nearest Neighbors Graph

The theory of subadditive and superadditive Euclidean functionals applies to one of the fundamental constructions of computational geometry, namely the k nearest neighbors graph. Such a graph puts an edge between each point in a set F and its k nearest neighbors. The length of the graph, denoted $N(k; F)$, is the sum of the lengths of these edges. The k nearest neighbors graph receives considerable attention in computational geometry and is used in classification problems as well as in the construction of algorithms for solving geometrical problems; see Preparata and Shamos (1985). In this section we describe the asymptotic behavior of $N(k; F)$ for random sets F.

We briefly review the known results describing the behavior of $N(k; U_1, ..., U_n)$, where U_i, $i \geq 1$, are i.i.d. uniform random variables on $[0, 1]^d$. To simplify the notation we will write $N(k; n)$ for $N(k; U_1, ..., U_n)$. Miles (1970) showed the following asymptotic result for $EN(k; n)$.

Theorem 8.1. *(Miles) The expected length of the k nearest neighbors graph $N(k; n)$ in dimension 2 satisfies*

$$\lim_{n \to \infty} \frac{EN(k; n)}{n^{1/2}} = C(k).$$

Later, Avram and Bertsimas (1993) used the technique of dependency graphs of Baldi and Rinott (1989) to show that $N(k; n)$ exhibits asymptotic normality. Their result adds to a similar result of Bickel and Breiman (1983), who show through long and complex arguments that $N(1; n)$ satisfies a central limit theorem (CLT). They were motivated to study the distribution of $N(1; n)$ in an attempt to derive the limiting distribution of a goodness of fit test for multidimensional densities based on nearest neighbor distances. Here $N(0, 1)$ denotes the standard normal random variable.

Theorem 8.2. *(CLT for k nearest neighbors) In dimension 2 $N(k; n)$ satisfies*

$$\lim_{n \to \infty} \frac{N(k; n) - EN(k; n)}{(\mathrm{Var} N(k; n))^{1/2}} \overset{\mathrm{d}}{=} N(0, 1).$$

This section will show that $N(k; n)$ is a smooth subadditive Euclidean functional of order 1 and in this way establish a law of large numbers which goes beyond Theorem 8.1. We are also able to treat the non-uniform case. Following McGivney (1997) we prove:

Theorem 8.3. *Let $X_1, ..., X_n$ be i.i.d. random variables with values in $[0, 1]^d$, $d \geq$ 2. Then*

$$(8.2) \qquad \lim_{n \to \infty} N(k; X_1, ..., X_n)/n^{(d-1)/d} = \alpha(k, d) \int_{[0,1]^d} f(x)^{(d-1)/d} \quad c.c.,$$

where f is the density of the absolutely continuous part of the law of X_1 and where $\alpha(k, d)$ is a constant depending only on k and d.

Proof of Theorem 8.3. We first show that $N(k; \cdot)$ is a smooth subadditive Euclidean functional. Given a rectangle R and a set F, we write $N(k; F, R)$ for $N(k, F \cap R)$ and in this way we turn $N(k, \cdot, \cdot)$ into a functional defined on the parameter set of rectangles. Here and throughout we adopt the convention that if $\text{card} F \leq k$, then the k nearest neighbors functional $N(k; F, R)$ allows points to have multiple matches to the same neighbor.

The k nearest neighbors functional is translation invariant and homogeneous of order 1. Moreover, we clearly have $N(k; F \cup G) \leq N(k; F) + N(k; G)$, that is $N(k; \cdot)$ is simply subadditive with no error term. It follows that the functional $N(k; \cdot, \cdot)$ satisfies geometric subadditivity (3.4) with no error term. Thus the nearest neighbors functional is a subadditive Euclidean functional of order 1 and thus satisfies the growth bounds of Lemma 3.3 with $p = 1$ there. We now show that it is smooth of order 1, that is we show

$$(8.3) \qquad |N(k; F \cup G) - N(k; F)| \leq C(\text{card} G)^{(d-1)/d}.$$

To establish (8.3) we call upon a simple lemma which shows that the vertices in the graph of the k nearest neighbors functional have bounded degree. A related lemma appears in Bickel and Breiman (1983).

Lemma 8.4. *Vertices in the k nearest neighbors graph have bounded degree.*

Proof. Consider the k nearest neighbors graph on $\{x_i\}_{i=1}^n$. We will show that the degree of x_n is finite and independent of n.

It is well-known (see e.g. Bickel and Breiman (1983)) that $\mathbb{R}^d$ can be expressed as the union of $C(d)$ disjoint cones having x_n as a common vertex. Were the degree of x_n to exceed $k \cdot C(d)$, then the pigeonhole principle would imply that at least one cone contains $k + 1$ points having x_n as a k nearest neighbor. Without loss of generality we may label these points $x_1, x_2, ..., x_{k+1}$ and we may assume that x_1 is farthest from x_n. However, by the definition of a cone, the distance between x_1 and the k points $x_2, ..., x_{k+1}$ is less than the distance between x_1 and x_n, showing that x_n is not one of the k nearest neighbors of x_1. Thus the degree of x_n is at most $k \cdot C(d)$. $\square$

We now show smoothness (8.3). By simple subadditivity we have

$$N(k; F \cup G) \leq N(k; F) + N(k; G) \leq N(k; F) + C(\text{card} G)^{(d-1)/d},$$

where the last inequality follows from the growth bounds. To complete the proof of smoothness (8.3) we need therefore only show

$$(8.4) \qquad N(k; F) \leq N(k; F \cup G) + C(\text{card}G)^{(d-1)/d}.$$

Given G, let F_G denote those points in F which have at least one of their k nearest neighbors in G. For any $H \subset F$, let $N'(k; H, F)$ denote the length of the k nearest neighbor graph on H with matching to points in F allowed. Note that $N'(k; H, F) \leq N'(k; F, F) = N(k, F)$ and $N'(k; H, F) \leq N(k; H)$. Thus

$$\begin{aligned} N(k; F) &= N(k; F, F) \\ &\leq N'(k; F - F_G, F) + N'(k; F_G, F) \\ &\leq N(k; F \cup G) + N(k; F_G). \end{aligned}$$

By Lemma 8.4, the cardinality of F_G is bounded by $C\text{card}G$, where $C := C(k, d)$. Thus by the growth bounds for $N(k; \cdot)$ we obtain

$$N(k; F_G) \leq C(\text{card}F_G)^{(d-1)/d},$$

proving (8.4) and establishing that the k nearest neighbors functional is smooth of order 1.

Having shown that the k nearest neighbors functional is a smooth subadditive Euclidean functional of order 1, we now consider the canonical *boundary* k nearest neighbors functional. Given a rectangle R and a finite subset $F \subset R$, the boundary k nearest neighbors functional is the length of the graph connecting each vertex $V \in F$ to its k nearest neighbors, where now points on the boundary of R are potential neighbors and eligible for consideration. V may be joined to a boundary point P up to k times and in general V is matched to P exactly j times, $1 \leq j \leq k$, if there are exactly $k - j$ points in F whose distance to V is less than the distance between P and V. Let $N_B(k; F, R)$ designate the boundary k nearest neighbors functional on $\mathcal{F} \times \mathcal{R}$.

It is straightforward to check that $N_B(k; F, R)$ is a smooth Euclidean functional of order 1, simply subadditive (2.2), and superadditive (3.3).

We now show that the k nearest neighbors functional is pointwise close (3.10) to its respective boundary functional, that is we show

$$(8.5) \qquad |N(k; F) - N_B(k; F)| = o\left((\text{card}F)^{(d-1)/d}\right).$$

This will be achieved by showing the more refined estimate

$$(8.6) \qquad N(k; F) \leq N_B(k; F) + C\left((\text{card}F)^{(d-2)/(d-1)} \vee \log(\text{card}F)\right).$$

To show (8.6) we need the analog of Lemma 3.8 and a little notation.

Lemma 8.5. *Let F be a subset of $[0,1]^d$ of cardinality n and consider the boundary k nearest neighbors graph on F. The sum S of the lengths of the edges connecting points in F with $\partial[0,1]^d$ is bounded by $C(k,d)\left(n^{(d-2)/(d-1)} \vee \log n\right)$.*

Proof. Follow the proof of Lemma 3.8 *verbatim* with $p=1$ and note that in each subcube Q of the partition $\mathcal{P}$ there are at most k points in $F \cap Q$ which are joined to the boundary. □

To show (8.6) we proceed as follows. Let $F_1 \subset F$ denote the subset of points which are linked to the boundary of $[0,1]^d$ by the boundary k nearest neighbors graph G_B on F. Let $\mathcal{B}$ denote the points on the boundary which are connected to F_1 by G_B. We note that

$$N(k;F) \le N(k;F_1) + N'(k;F-F_1,F),$$

where the second term on the right side is bounded by $N_B(k;F)$. To prove (8.5) it will suffice to show that if $n := \text{card}F$ then

(8.7) $N(k;F_1) \le C(n^{(d-2)/(d-1)} \vee \log n).$

To show (8.7), consider the optimal tour on $\mathcal{B}$. Clearly $\text{card}\mathcal{B} \le n$, and so

$$T(\mathcal{B}) \le Cn^{(d-2)/(d-1)},$$

since $\mathcal{B}$ lies in a set of dimension $d-1$. We use the graph T given by $T(\mathcal{B})$ to construct a k neighbors graph on F_1. Let V_i, $i \ge 1$, be an enumeration of the points in F_1. Without loss of generality we may assume that $\text{card}F_1 \ge k+1$. We will assume that $N_B(k;F)$ connects V_i to $B_i \in \mathcal{B}$ for all choices of i. Relabeling if necessary we may without loss of generality let $B_1, B_2, ..., B_k, B_{k+1}, ...$ be the successive points in $\mathcal{B}$ visited by the tour T. Consider V_1. Choose the k neighbors of V_1 to be $V_2, ..., V_{k+1}$. The edges joining V_1 to these newly defined neighbors $V_2, ..., V_{k+1}$ come at a cost which is bounded by the sum of $\sum_{i=2}^{k+1} \|B_i - V_i\|$, $k\|B_1 - V_1\|$, and $\sum_{i=2}^{k+1} \|B_1 - B_i\|$. This latter sum is in turn bounded by a constant multiple $C(k)$ of the sum $\sum_{i=1}^{k} \|B_i - B_{i+1}\|$.

Now repeat the above construction for the remaining points in F_1. For each point V_i in F_1 we find thus k neighbors in F_1; they are not necessarily the nearest neighbors. The total cost of performing this operation is bounded by $C(k)(S + T(\mathcal{B}))$. The previously announced bounds on S and $T(\mathcal{B})$ now give the required result (8.7). This completes the proof of pointwise closeness and proves Theorem 8.3. □

Remark. We may similarly establish that the k nearest neighbors functional is close in mean (3.16) to the boundary functional. Take $F := \{U_1, ..., U_n\}$ in the above analysis and show that the sum $S(n)$ of the lengths of the edges connecting points in F with the boundary satisfies $E(S(n)) \le Cn^{(d-2)/d}$. We also apply the bound $E\text{card}\mathcal{B} \le Cn^{(d-1)/d}$, which may be established using the methods for proving Lemma 3.10.

8.4.   The Many Traveling Salesman Problem

We consider a generalization of the TSP in which more than one salesman (and thus more than one tour) is allowed. In the k-traveling salesman problem on F (the k-TSP) we find a collection C of k subtours, each containing a distinguished vertex V, such that each point in F is in one subtour. $T(k; C, F)$ denotes the sum of the combined lengths of the k subtours in C and we define the k-TSP functional as the infimum $T(k; F) := \inf_C T(k; C, F)$. This problem models the situation in which k salesmen work for a company with a home office at V and between them they need to visit each city once.

Figure 8.3.   The many traveling salesman problem ($k = 3$)

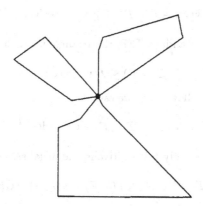

A similar version of the k-TSP involves minimizing the maximum length subtour in the collection of subtours on F. This problem has received great attention but we will not consider it here. We refer to Christofides et al. (1979, Chapter 11) for a general discussion.

The k-TSP resembles the distribution problem in which k vehicles based at a central facility or depot are required to visit geographically dispersed customers in order to distribute or collect commodities. The collection and delivery of mail from mail boxes is one example of what is generally termed the vehicle routing problem (VRP). The k-TSP on F involves finding the shortest overall length of k subtours which originate at the depot and pass through all points of F. There is however no distribution or collection of commodities.

Observe that the k-TSP is not in general homogeneous nor does it satisfy translation invariance $T(F + y) = T(F)$ if $V \notin F$. These limitations are easily overcome if we view the k-TSP as a functional defined over d-dimensional rectangles, a point of view which has been the common thread throughout this monograph.

We now describe the k-TSP *functional* over the d-dimensional rectangles. This is a natural context for the k-TSP and brings out its intrinsic subadditivity. Given a rectangle R in $(\mathbb{R}^+)^d$, let $V := V(R)$ denote the vertex which is closest to the origin.

If $F \subset R$ is a finite set, then we let $T(k; F, R)$ denote the k-TSP tour length $T(k; F)$, where each of the k subtours must contain the distinguished vertex V. $T(k; F, R)$ is a functional indexed by the d-dimensional rectangles and is subadditive in the sense of (3.4), i.e.,

$$T(k; F, R) \leq T(k; F \cap R_1, R_1) + T(k; F \cap R_2, R_2) + C \operatorname{diam} R.$$

To see this, consider the $2k$ subtours on R_1 and R_2. These subtours pass through the vertices $V(R_1)$ and $V(R_2)$. By the triangle inequality, the $2k$ tours on R_1 and R_2 may be tied together to generate k subtours on R at an extra cost of at most $Ck(\operatorname{diam} R)$. These k subtours each pass through $V(R)$ and generate a feasible solution of the k-TSP on R. Thus subadditivity follows.

We now verify that the k-TSP is smooth of order 1, that is

$$(8.8) \qquad\qquad |T(k; F) - T(k; F \cup G)| \leq C (\operatorname{card} G)^{(d-1)/d}.$$

As is the case with the standard TSP, the triangle inequality shows that $T(k; \cdot)$ is monotone so that

$$T(k; F) \leq T(k; F \cup G).$$

Smoothness of order 1 follows once we establish

$$(8.9) \qquad\qquad T(k; F \cup G) \leq T(k; F) + C (\operatorname{card} G)^{(d-1)/d}.$$

However, the k-TSP is simply subadditive in the usual sense

$$T(k; F \cup G) \leq T(k; F) + T(k; G) + Ck$$

since we may obtain a k tour graph on $F \cup G$ by tying together the k tour graph on F with the k tour graph on G at an additional cost of Ck. Since $T(k; G) \leq C (\operatorname{card} G)^{(d-1)/d}$, we get (8.9) and thus smoothness (8.8).

The boundary k-TSP functional on rectangles R is defined in a natural way. Given $F \subset R$, the graph of a feasible boundary k-TSP tour consists of at most k subtours joined to the distinguished vertex $V(R)$. Moreover, subtours may exit to the boundary of R and subsequently re-enter. Travel along the boundary is free. The minimal total edge length of such a graph is called the boundary k-TSP functional and is denoted by $T_B(k; F, R)$; it is the natural analog of the canonical boundary TSP functional $T_B(F, R)$ studied in earlier chapters.

$T_B(k; F, R)$ is superadditive in the sense of (3.3), that is

$$T_B(k; F, R) \geq T_B(k; F \cap R_1, R_1) + T_B(k; F \cap R_2, R_2),$$

where $R := R_1 \cup R_2$ is the union of rectangles. To see this, observe that the restriction of the graph given by $T_B(k; F, R)$ to the subrectangle R_i, $1 \leq i \leq 2$, consists of a collection of paths with endpoints lying on the boundary of R_i. These paths may be linked with edges lying on the boundary of R_i, $1 \leq i \leq 2$, to form a collection of at most k closed loops, each of which passes through the distinguished vertex $V := V(R_i), 1 \leq i \leq 2$. Since the extra edges used in this operation all lie on the boundary of R_i, this operation comes at no additional cost. The resulting collection of at most k subtours can only exceed the minimal such collection, showing that the restriction of the graph given by $T_B(k; F, R)$ to R_i must exceed $T_B(k; F \cap R_i, R_i)$, $1 \leq i \leq 2$. This establishes superadditivity.

We may prove that $T_B(k; F, R)$ is smooth of order 1 by following the smoothness proof for $T(k; F, R)$.

Finally, the k-TSP functional is both pointwise close (3.10) and close in mean (3.15) to the canonical boundary k-TSP functional. This is seen by following the proofs of Lemmas 3.7 and 3.10; no new ideas are needed.

We have thus established that the k-TSP problem is a subadditive Euclidean functional which is smooth of order 1 and that the canonical k-TSP boundary functional is superadditive and smooth of order 1. Since they are pointwise close and close in mean they satisfy the basic limit Theorem 4.1, the rate results of Chapter 5, and the umbrella Theorem 7.1.

8.5. The Greedy Matching Heuristic

The theory of subadditive and superadditive Euclidean functionals evidently covers a wide range of problems in geometric probability. In the sequel we will see that the triangulation problem and the k-median problem of Steinhaus also fit neatly into the present theory. Yet the theory doesn't cover all problems in combinatorial optimization and operations research. Indeed, some heuristic solutions unfortunately do not seem to fit neatly into the theory. In this section we will see that the greedy heuristic for minimal matching does not quite fit into our theory.

The greedy heuristic provides an algorithm for obtaining a Euclidean matching which closely approximates the optimal matching. The greedy heuristic $G(F)$ on a point set F successively matches the closest unmatched pairs of points. Much work has been done to show that the heuristic G is a good approximation to the minimal matching S. Reingold and Tarjan (1981) have shown that for point sets $F \in \mathbb{R}^2$ of cardinality n, one has the bound

$$\frac{G(F)}{S(F)} \leq \frac{4}{3} n^{\log_2 1.5}.$$

The greedy heuristic G is clearly translation invariant and homogeneous. Less obvious is the fact that G is smooth of order 1, a fact proved by Avis, Davis, and Steele (1988). The proof of smoothness depends in part on a growth estimate $G(F) \leq C(\text{card} F)^{(d-1)/d}$. Thus the heuristic G is a smooth Euclidean functional of order 1. However, examples show that G is apparently not subadditive in the usual sense (3.4). The canonically defined boundary functional G_B is apparently not superadditive either.

Despite the fact that the greedy heuristic and its canonically defined boundary functional are not subadditive and superadditive, respectively, Avis, Davis, and

Steele (1988) showed that the heuristic can be approximately localized by the sum of its values on the m^d subcubes of $[0,1]^d$. In this way they showed that the greedy heuristic exhibits the asymptotic behavior (7.2) given by the umbrella Theorem 7.1.

This raises the following question: can the theory of smooth subadditive Euclidean functionals be appropriately modified and enlarged to accommodate functionals such as the greedy matching heuristic which are Euclidean and smooth but not subadditive in the sense of (3.4)? We believe that this question can be resolved positively. Ideally we would like to enlarge the theory so that it also encompasses other heuristics such as the greedy heuristic for the TSP. It is not clear that this can be done, however.

8.6.   The Directed TSP

Consider the random directed graph G_n whose vertices are independent and uniformly distributed random variables $U_1, ..., U_n$ on the unit square. For $1 \leq i < j \leq n$, the orientation of the edge $X_i X_j$ is selected at random, independently for each edge and independently of the U_i, $i \geq 1$. Thus the edges in G_n are given a direction just by flipping fair coins. The directed TSP involves finding the shortest directed path through the random vertex set. While it is clear that there may not be a directed cycle through the vertex set, a classic result of Rédei (1934) nonetheless guarantees that a path exists. We write $D(n) := D(U_1, ..., U_n)$ for the length of the directed TSP path through the sample $U_1, ..., U_n$.

While the directed TSP is Euclidean and subadditive, it is not clear that it is smooth. Nonetheless, Steele (1986) has shown that $D(n)$ behaves as though it were a subadditive Euclidean functional which is smooth of order 1:

$$\lim_{n \to \infty} ED(n)/n^{1/2} = \alpha.$$

Steele's (1986) methods do not yield the complete convergence of the directed TSP and it was left to Talagrand (1991) to make this improvement. Using martingale difference sequences and an inequality related to Azuma's inequality, Talagrand (1991) actually shows the following stronger result, which implies the complete convergence of the directed TSP:

$$\sum_{n=1}^{\infty} P\{|D(n) - ED(n)| > C \log n\} < \infty.$$

As with any application of Azuma's inequality, the hard part is to obtain sharp bounds on the martingale difference sequence associated to $D(n)$. Talagrand (1991) handles this by showing that for t in the range $C(\log n/n)^{1/2} \leq t \leq 1$ we have

$$P\{|D(n) - D(n+1)| \geq t\} \leq C \exp(-t^2 n/C).$$

It is unresolved whether the directed TSP satisfies the asymptotics (7.2) of the umbrella Theorem 7.1.

Notes and References

1. In addition to the k nearest neighbors functional, there are other problems of computational geometry which have some or all of the properties of smooth subadditive Euclidean functionals. We mention the total edge length of the Voronoi tessellation, Gabriel graph, and Delaunay triangulation. Concerning the first two, Talagrand (1995, Chapter 11) shows that these constructions satisfy a smoothness condition.

Let $V(X_1, ..., X_n)$ denote the total edge length of the Voronoi tessellation on random variables $X_1, ..., X_n$. If $X_1, X_2, ...$ are independent and have a common density $f(x)$ on the unit square which satisfies $0 < \alpha \leq f(x) \leq \gamma$ for constants α and γ then McGivney and Yukich (1997a) show that $\lim_{n \to \infty} V(X_1, ..., X_n)/n^{1/2} = 2 \int_{[0,1]^2} f(x)^{1/2} dx$ c.c. This extends upon Miles (1970).

2. We expect that the work of Jaillet (1993c) will also fit into the theory of subadditive Euclidean functionals. Jaillet considers a probabilistic version of the TSP in the following sense: find the shortest tour T through a point set V of cardinality n. For any given instance of the problem, only a random subset V' of points from V has to be visited. The subset V' is visited in the *same order* as they appear in the tour T; this gives a tour T' through V'. The length of the tour T' is a Euclidean functional and it seems likely that it may satisfy the umbrella Theorem 7.1.

3. We also expect that the degree-K minimum spanning tree (MST) problem will also fit into our theory. The degree-K MST problem asks for the minimum length spanning tree that has no vertex of degree greater than K. Such problems have been studied by Papadimitriou and Vazirani (1984) among others.

4. There are several well-known heuristics for the TSP and in his survey Steele (1990b) asks whether they satisfy the asymptotics (7.2). Goemans and Bertsimas (1991) show that the Held-Karp (1970,1971) heuristic enjoys the limit result (7.2). Although they do not use methods involving boundary functionals it seems likely that their results could be handled (and therefore extended) by the approach discussed in this monograph. There are other heuristics for the TSP and we refer to Lawler et al. (1985), Rosenkrantz et al. (1977), and Papadimitriou and Steiglitz (1982). Some of these heuristics may satisfy the umbrella Theorem 7.1. In particular, it is not yet clear whether the 2-OPT heuristic and the Christofides algorithm for the TSP satisfy the asymptotics (7.2).

5. Semi-matchings. One could easily define a power-weighted edge version of $\hat{S}$ but we do not consider this generalization here. We note that the solution to the semi-matching problem runs in polynomial time.

6. For additional limit results for the k nearest neighbors graph with power-weighted edges we refer to McElroy (1997) and McGivney (1997).

9. MINIMAL TRIANGULATIONS

9.1. Introduction

This chapter examines the asymptotic behavior of the total edge length of the minimal triangulation of n points which are independently and identically distributed on the unit square.

Minimal triangulations arise naturally in many areas of mathematics and have particular importance in computational geometry. They have applications to surface interpolation, geometric searching techniques, and the finite-element method. See Preparata and Shamos (1985) and Bern and Eppstein (1992) for thorough treatments. Steele (1982) took the first important steps in the study of the triangulation functional on random points in the unit square.

We now formulate the minimal triangulation problem precisely. Given a finite set F in $[0,1]^2$, a triangulation of F is a decomposition of $[0,1]^2$ into triangles whose vertices coincide with F and the four corners of $[0,1]^2$. In general, a set F admits more than one triangulation, which need not be a simplex. The *total edge length* of a triangulation is the sum of the lengths of the edges in the triangulation. See Figure 9.1.

Figure 9.1. A triangulation of a three point set

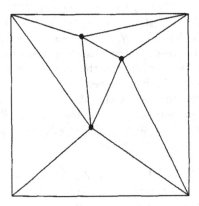

Given $1 < \delta < \infty$, a "δ-triangulation" of F is a triangulation in which all triangles have aspect ratios which are less than δ, that is for all triangles the ratio of the radii of the circumscribed ball to the inscribed ball is less than δ. Such triangulations have a number of applications and motivations, including some in computational learning theory (Salzberg et al., 1991). Let S_δ denote a function which assigns to each set $F \subset [0,1]^2$ a δ-triangulation of F which has the least total edge length. Let $S_\delta(F)$ denote the graph of this triangulation (possibly empty depending on the choice of F and δ) and let $|S_\delta(F)|$ denote the total edge length of $S_\delta(F)$. We will occasionally write $S(F)$ for $S_\delta(F)$.

Figure 9.2. A triangulation of collinear points (no Steiner points allowed)

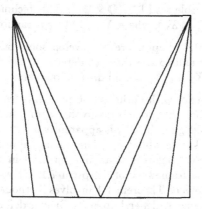

Figure 9.3. Adding a Steiner point P decreases total edge length

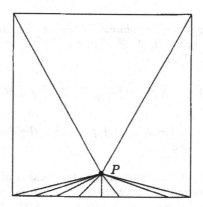

If $G \subset [0,1]^2$ is an additional set of points distinct from F (i.e., a Steiner set), then $S_\delta(F \cup G)$ denotes the graph of a *Steiner δ-triangulation* of F. We tacitly assume throughout and without further mention that $\delta > 1$ is chosen large enough such that for every set F there is a Steiner set G for which $S_\delta(F \cup G)$ exists. We define the *length of the minimal Steiner δ-triangulation* of F by

$$\sigma_\delta(F) := \inf_G |S_\delta(F \cup G)|,$$

where G ranges over all finite sets of Steiner points including the empty set. It is conceivable that adding more and more Steiner points may decrease the total length of the triangulation (see Bern and Eppstein (1992)). Thus is it unclear whether the infimum is realized by a set G and it is thus an open problem whether a "minimal δ-triangulation" actually exists. See Figures 9.2 and 9.3.

Without loss of generality we assume that G ranges over points in $[0,1]^2$ with rational coordinates. The set of admissible Steiner points of a set F thus has the cardinality of the countable set $\bigcup_{n=1}^{\infty}(\mathbb{Q} \times \mathbb{Q})^n$. This technical remark ensures the measurability of $\sigma_\delta(X_1,...,X_n)$, where X_i, $i \geq 1$, are random variables.

The main goals of this chapter are to develop some basic deterministic properties of triangulations and to use these to determine the asymptotic behavior of $\sigma_\delta(X_1,...,X_n)$, where X_i, $i \geq 1$, are i.i.d. random variables.

Based on our experience with optimization problems, we would expect that the minimal triangulation length σ_δ conforms to the asymptotics of Theorem 7.1. The proof is a bit more challenging since σ_δ is apparently not "simply subadditive" (2.2) and since not much is known about σ_δ. The main goal of this chapter is develop some basic properties of stochastic triangulations and in this way show that σ_δ fits naturally within the framework of Theorem 7.1. Showing that triangulations conform to the conditions of Theorem 7.1 involves methods combining probability and geometry. The arguments depend upon a judicious definition of a superadditive "boundary triangulation functional". By verifying that subadditive triangulations and superadditive boundary triangulations satisfy the conditions of Theorem 7.1 when $p = 1$ and $d = 2$, we prove:

Theorem 9.1. *(asymptotics for minimal triangulations) Let X_i, $i \geq 1$, be i.i.d. random variables with values in $[0,1]^2$. Fix $1 < \delta < \infty$ and consider the minimal triangulation length σ_δ. Then*

$$(9.1) \qquad \lim_{n \to \infty} \sigma_\delta(X_1,...,X_n)/n^{1/2} = \alpha(\sigma_\delta)\int_{[0,1]^2} f(x)^{1/2}dx \quad c.c.,$$

where $\alpha(\sigma_\delta)$ is a positive constant and f denotes the density of the absolutely continuous part of the law of X_1.

By placing the triangulation functional in the context of Euclidean functionals, we may moreover derive rates of convergence in a natural way. It is not clear how much improvement can be made in these rates. Let $U_1, U_2, ...$ be i.i.d. with the uniform distribution on $[0,1]^2$.

Theorem 9.2. *Fix $1 < \delta < \infty$. The mean of σ_δ satisfies*

$$(9.2) \qquad |E\sigma_\delta(U_1,...,U_n) - \alpha(\sigma_\delta)n^{1/2}| \leq C(n\log n)^{1/4},$$

where $C := C(\delta)$ is a constant depending only on δ.

The above two-dimensional results have a natural three-dimensional analog. Given a finite set F in $[0,1]^3$, a tetrahedralization of F is a decomposition of $[0,1]^3$ into tetrahedra whose vertices coincide with the points in F and the corners of the cube $[0,1]^3$. In general, F admits more than one tetrahedralization. The *total surface area* of a tetrahedralization is the sum of the areas of the triangular faces.

Given $1 < D < \infty$, a D-tetrahedralization is one in which the tetrahedra have aspect ratios less than D, that is the ratio of the radii of the circumscribed sphere to the inscribed sphere is less than D for all tetrahedra. This regularity condition insures that the cube of the length of a tetrahedral edge is bounded by a constant multiple of the volume of the tetrahedron, a fact which will be useful in the sequel.

Given $1 < D < \infty$, let T_D denote a function which assigns to each set $F \subset [0,1]^3$ a D-tetrahedralization of F having the least total surface area. Let $T_D(F)$ denote the graph of the tetrahedralization and let its total surface area be denoted by $|T_D(F)|$. If $G \subset [0,1]^3$ is a Steiner set, then we let $T_D(F \cup G)$ denote the graph of the corresponding Steiner D-tetrahedralization of F. We tacitly assume that $D > 1$ is chosen large enough so that for every set $F \subset [0,1]^3$, there is a Steiner set G for which $T_D(F \cup G)$ exists. Analogously to $\sigma_\delta(F)$, define the area $\tau_D(F)$ of the *minimal* Steiner D-tetrahedralization of F by

$$\tau_D(F) := \inf_G |T_D(F \cup G)|,$$

where G ranges over all Steiner sets. $\tau_D(F)$ may be thought of as the *discrete Plateau functional* for the point set F.

As in the definition of σ_δ, we may without loss of generality restrict attention to Steiner points with rational coordinates. This ensures the measurability of $\tau_D(X_1, ..., X_n)$, where X_i, $i \geq 1$, are random variables.

In section 9.6 we will show that the Steiner tetrahedralization functional satisfies the conditions of Theorem 7.1 with $p = 2$ and $d = 3$. Notice that the order is 2 since τ_D involves sums of surface areas. Steiner tetrahedralizations furnish a natural example of a functional in geometric probability which has an order larger than 1. By applying Theorem 7.1 we will prove the following analog of Theorem 9.1. This makes progress on a question raised by Beardwood, Halton, and Hammersley (1959).

Theorem 9.3. *(asymptotics for minimal tetrahedralizations) Let X_i, $i \geq 1$, be i.i.d. random variables with values in $[0,1]^3$. Then for each fixed $1 < D < \infty$ we have*

$$(9.3) \qquad \lim_{n \to \infty} \tau_D(X_1, ..., X_n)/n^{1/3} = \alpha(\tau_D) \int_{[0,1]^3} f(x)^{1/3} dx \quad c.c.,$$

where $\alpha(\tau_D)$ is a positive constant and where f denotes the density of the absolutely continuous part of the law of X_1.

We anticipate that Theorems 9.1 and 9.3 admit extensions to higher dimensions. In this way we could perhaps find asymptotics for the randomized version of the problem of Douglas (1939) which considers minimal surfaces in higher dimensions. Beardwood, Halton, and Hammersley (1959) were apparently the first to consider such a problem. In their somewhat cryptic remarks they recognize the potential applicability of subadditivity methods but do not develop the necessary mathematics. For a full treatment of the problem of Douglas we refer to Courant and Schiffer (1950, Chapter 4) and Douglas (1939). We will not consider generalizations to higher dimensions in this monograph.

To facilitate the exposition and lighten the notation, we will henceforth omit mention of δ when referring to δ-triangulations. Moreover, when it is clear from the context, we will often simply write σ for σ_δ and τ for τ_D.

9.2.   The Boundary Triangulation Functional

In the previous chapters "boundary functionals" played an important role in establishing the intrinsic superadditivity of problems in Euclidean optimization. The purpose of this section is to appropriately define the "boundary δ-triangulation functional" and to use it to prove Theorems 9.1 and 9.2.

Given any subset $F \subset \mathbb{R}^2$ whose convex hull co(F) contains $[0,1]^2$, consider a δ-triangulation of co(F). Such a δ-triangulation partitions co(F) into triangles whose vertices coincide with F and whose aspect ratios are bounded by δ. A "boundary δ-triangulation" of $[0,1]^2$ with respect to F, denoted here by $S_{B,\delta}(F,[0,1]^2)$, is obtained by considering the intersection of $[0,1]^2$ and the graph of a δ-triangulation of co(F). A boundary δ-triangulation of $[0,1]^2$ thus partitions $[0,1]^2$ into the usual triangles as well as perhaps quadrilaterals, pentagons, and even hexagons. When the context is clear, we will omit mention of δ and refer to boundary δ-triangulations as simply boundary triangulations.

By the "total edge length" $|S_{B,\delta}(F,[0,1]^2)|$ of the boundary δ-triangulation $S_{B,\delta}(F,[0,1]^2)$ we mean the sum of the lengths of the edges in the triangulation which lie in the interior of $[0,1]^2$. Analogously to $\sigma(F) := \sigma_\delta(F)$, define for all $F \subset [0,1]^2$ the length of the "minimal boundary δ-triangulation" of F with respect to $[0,1]^2$ by

$$\sigma_{B,\delta}(F,[0,1]^2) := \inf_G |S_{B,\delta}(F \cup G,[0,1]^2)|,$$

where G ranges over finite sets of Steiner points with the property that the convex hull of $F \cup G$ contains $[0,1]^2$. Without loss of generality we will assume that the points in G have rational coordinates. Minimal boundary triangulations may fail to exist for the same reasons that the standard minimal triangulation may not exist. We will occasionally condense notation and write σ_B for $\sigma_{B,\delta}$ and S_B for $S_{B,\delta}$. It is clear from the definitions that $\sigma_B(F,[0,1]^2) \leq \sigma(F)$ for all $F \subset [0,1]^2$.

Given any convex polygon Δ and a point set F such that co$(F) \supset \Delta$, we extend the above definitions and define a boundary δ-triangulation of Δ with respect to F as the intersection of Δ and the graph of a δ-triangulation of co(F). When $F \subset [0,1]^2$, $\sigma_B(F,\Delta) := \sigma_{B,\delta}(F,\Delta)$ denotes the length of the minimal boundary δ-triangulation of F with respect to Δ and is defined analogously to $\sigma_B(F,[0,1]^2)$; notice that $\sigma_B(F,\Delta)$ does not count the lengths of edges lying on the boundary of Δ.

The minimal boundary triangulation length σ_B enjoys geometric superadditivity (3.3): for all $F \subset [0,1]^2$ we have

$$(9.4) \qquad \sigma_B(F, [0,1]^2) \geq \sum_{i=1}^{m^2} \sigma_B(F \cap Q_i, Q_i),$$

where Q_i, $1 \leq i \leq m^2$, denotes the usual partition of $[0,1]^2$ into subsquares of edge length m^{-1}.

To see this, find a sequence G_n of Steiner sets such that the boundary triangulations $S_B(F \cup G_n, [0,1]^2)$, $n \geq 1$, have lengths $|S_B(F \cup G_n, [0,1]^2)|$ decreasing down to $\sigma_B(F, [0,1]^2)$ as n goes to infinity. For each $1 \leq i \leq m^2$, let $S_B^i(F \cup G_n, [0,1]^2)$ denote the boundary triangulation of Q_i generated by the intersection of $S_B(F \cup G_n, [0,1]^2)$ with subsquare Q_i. Then for each $n \geq 1$ we have

$$|S_B(F \cup G_n, [0,1]^2)| \geq \sum_{i=1}^{m^2} |S_B^i(F \cup G_n, [0,1]^2)| \geq \sum_{i=1}^{m^2} \sigma_B(F \cap Q_i, Q_i),$$

where the last inequality follows by minimality of σ_B. Let n tend to infinity to deduce superadditivity (9.4). Summarizing, we have shown:

Lemma 9.4. σ_B *is superadditive.*

Having defined the minimal triangulation lengths σ and σ_B we are positioned to prove Theorem 9.1. We must verify that σ and σ_B are smooth subadditive and superadditive Euclidean functionals of order 1, respectively, and that they satisfy the closeness condition (3.15) with $p = 1$ and $d = 2$. This is shown in the remainder of the chapter.

9.3. Minimal Triangulations are Subadditive and Smooth

We will verify that the length $\sigma := \sigma_\delta$ of the minimal triangulation is a subadditive Euclidean functional of order 1. Throughout, let $1 < \delta < \infty$ be arbitrary but fixed and write σ for σ_δ. In the sequel we show that the boundary triangulation length $\sigma_B := \sigma_{B,\delta}$ is a superadditive Euclidean functional of order 1.

It will be helpful to consider the triangulation of a set in a region other than the unit square. We thus enlarge the definition of triangulations in the following way.

Definition 9.5. *Let $F \subset \Delta$ be a finite set, where $\Delta \subset [0,1]^2$ is a convex polygon. Consider all Steiner δ-triangulations $S_\delta(F \cup G, \Delta)$, $G \subset \Delta$, of Δ, i.e. all decompositions of Δ into triangles whose vertices coincide with $F \cup G$ and the corners of Δ and whose aspect ratios are bounded by δ. Let $S_\delta(F \cup G, \Delta)$ have total edge length $|S_\delta(F \cup G, \Delta)|$. Define*

$$\sigma_\delta(F, \Delta) := \inf_G |S_\delta(F \cup G, \Delta)|,$$

where G ranges over all finite sets in Δ. We call $\sigma(F, \Delta) := \sigma_\delta(F, \Delta)$ the length of the minimal δ-triangulation of F with respect Δ. When $\Delta = [0,1]^2$, we will simply write $\sigma(F)$ for $\sigma(F, [0,1]^2)$.

Notice that σ, considered as a function on pairs (F, R) is Euclidean. σ also satisfies geometric subadditivity (3.5) with no error term:

Lemma 9.6. (*subadditivity of minimal triangulations*) *For every* $F \subset [0,1]^2$ *we have*

$$(9.5) \qquad \sigma(F) \le \sum_{i=1}^{m^2} \sigma(F \cap Q_i, Q_i).$$

Proof. For each $1 \le i \le m^2$, find a sequence of Steiner sets $G_n := G_{n,i} \subset Q_i$, $n \ge 1$, such that the triangulations $S_\delta((F \cup G_n) \cap Q_i, Q_i)$, $n \ge 1$, have lengths $|S_\delta((F \cup G_n) \cap Q_i, Q_i)|$ which decrease down to the length of the minimal triangulation $\sigma(F \cap Q_i, Q_i)$ as n tends to infinity. For each $n \ge 1$, the union of the local triangulations $S_\delta((F \cup G_n) \cap Q_i, Q_i)$, $1 \le i \le m^2$, is a feasible Steiner triangulation of $[0,1]^2$. Minimality implies that for all $n \ge 1$

$$\sigma(F) \le \sum_{i=1}^{m^2} |S_\delta((F \cup G_n) \cap Q_i, Q_i)|.$$

Now let n tend to infinity to deduce (9.5). $\quad\square$

When Q_i and Q_j are adjacent subsquares the Steiner points on the boundary of Q_i need not coincide with the Steiner points on the boundary of Q_j. We notice therefore that (9.5) would fail if we restricted attention to triangulations which were simplicial complexes.

Geometric subadditivity implies that σ satisfies the growth bounds of Lemma 3.3. Moreover, we may obtain growth bounds for $\sigma_\delta(F, \Delta)$, $\Delta \subset [0,1]^2$ a convex polygon, by approximating Δ by the union of inscribed subsquares and applying growth bounds on the individual subsquares. This argument, whose details are left to the reader, shows:

Lemma 9.7. *There is a finite constant C such that for all convex polygons $\Delta \subset [0,1]^2$ and all non-empty sets $F \subset \Delta$*

$$(9.6) \qquad \sigma_\delta(F, \Delta) \le C(\mathrm{card}F)^{1/2}.$$

Finally, we may verify that σ is smooth of order 1, namely

$$(9.7) \qquad |\sigma(F \cup G) - \sigma(F)| \le C(\mathrm{card}G)^{1/2}.$$

For all $\varepsilon > 0$ and all $F \subset [0,1]^2$ let $G_\varepsilon := G_\varepsilon(F) \subset [0,1]^2$ be (uniquely defined) Steiner sets such that

$$(9.8) \qquad |S_\delta(F \cup G_\varepsilon)| \le \sigma_\delta(F) + \varepsilon.$$

Here and henceforth, let $\Delta^\varepsilon(F)$ denote the collection of triangles defined by the triangulation $S_\delta(F \cup G_\varepsilon)$ and let $\mathcal{E}^\varepsilon := \mathcal{E}^\varepsilon(F)$ denote the collection of edges of these triangles. By assumption, the aspect ratios of the triangles in $\Delta^\varepsilon(F)$ are uniformly bounded by δ. It follows that the square of the length of an edge of a triangle is bounded by a constant multiple of the triangular area. Since the sum of the areas of the triangles in $\Delta^\varepsilon(F)$ is just the area of the unit square, it follows that

$$(9.9) \qquad \sum_{E \in \mathcal{E}^\bullet} |E|^2 \leq C$$

for some universal constant $C := C(\delta)$ which doesn't depend on ε. Here and elsewhere, $|E|$ denotes the Euclidean length of the edge E.

To show smoothness, it suffices by (9.9) to show

$$(9.10) \qquad \sigma(F) \leq \sigma(F \cup G) \leq \sigma(F) + C \left(\sum_{E \in \mathcal{E}^\bullet} |E|^2 \right)^{1/2} (\mathrm{card} G)^{1/2}.$$

Notice that the first inequality in (9.10) is a consequence of the intrinsic monotonicity of σ. To show (9.7), it thus suffices to prove the second inequality in (9.10).

Given G, we may assume that $G \cap F = \emptyset$. The points in G are located in triangles $\Delta_1, ..., \Delta_J$, $J := J(G) < \infty$, belonging to $\Delta^\varepsilon(F)$ (if a point in G lies on an edge in $\mathcal{E}^\varepsilon(F)$, then it belongs to *two* triangles in $\Delta^\varepsilon(F)$). Let E_i, $1 \leq i \leq J$, be the longest edge of triangle Δ_i, $1 \leq i \leq J$.

Observe that $\sigma(F \cup G)$ is bounded by $|S_\delta(F \cup G_\varepsilon)|$ and the sum of the lengths of the minimal triangulations of $G \cap \Delta_i$ with respect to Δ_i, $1 \leq i \leq J$. In other words

$$\sigma(F \cup G) \leq |S_\delta(F \cup G_\varepsilon)| + \sum_{i=1}^{J} \sigma(G \cap \Delta_i, \Delta_i)$$

$$(9.11) \qquad \leq |S_\delta(F \cup G_\varepsilon)| + C \sum_{i=1}^{J} |E_i| \, (\mathrm{card}(G \cap \Delta_i))^{1/2}$$

by scaling and Lemma 9.7. Hölder's inequality implies

$$\sigma(F \cup G) \leq |S_\delta(F \cup G_\varepsilon)| + C \left(\sum_{E \in \mathcal{E}^\bullet} |E|^2 \right)^{1/2} (\mathrm{card} G)^{1/2},$$

which together with (9.8)–(9.9) gives smoothness (9.7) as desired.

9.4. Boundary Minimal Triangulations

By Lemma 9.4 we know that the boundary triangulation functional $\sigma_B := \sigma_{B,\delta}$ is superadditive where $1 < \delta < \infty$ is arbitrary but fixed. It is also clear that σ_B is Euclidean. It remains to verify smoothness

$$(9.12) \qquad |\sigma_B(F \cup G, [0,1]^2) - \sigma_B(F, [0,1]^2)| \leq C(\mathrm{card} G)^{1/2}.$$

This will follow from a slight modification of the proof of smoothness of σ.

We first clarify the terminology. Let $1 < \delta < \infty$ be arbitrary but fixed. As before, for all $\varepsilon > 0$ and all $F \subset [0,1]^2$, let $G_\varepsilon := G_\varepsilon(F) \subset \mathbb{R}^2$ be (uniquely defined) Steiner sets such that

$$|S_B(F \cup G_\varepsilon)| \leq \sigma_B(F) + \varepsilon.$$

Without loss of generality we may assume that $\mathrm{co}(F \cup G_\varepsilon)$ is contained in a large square $Q \supset [0,1]^2$ where the edge length of Q is at most $C := C(\delta)$.

Let $\Delta_B^\varepsilon(F)$ denote the collection of polygons formed by the boundary triangulation $S_B(F \cup G_\varepsilon)$. Let $\mathcal{E}_B^\varepsilon := \mathcal{E}_B^\varepsilon(F)$ denote the set of all edges of these polygons. Since these edges form a subset of the edges of triangles contained in the square Q, it follows as in (9.9) that the sum of the squares of their lengths is bounded by a constant multiple of the area of Q, that is

$$\sum_{E \in \mathcal{E}_B^\varepsilon} |E|^2 \leq C$$

where $C := C(\delta)$. If Δ is a polygon in $\Delta_B^\varepsilon(F)$ with diameter D and if G is a set of points in Δ then $\sigma(G, \Delta)$ is bounded by $C \cdot D \cdot (\mathrm{card}G)^{1/2}$. D is bounded by the sum of the lengths of the edges of Δ and D^2 is bounded by a constant multiple of the sum of the squares of the lengths of the edges of Δ. The proof of smoothness (9.12) follows exactly as in the proof of the smoothness (9.7) of σ.

9.5. Closeness in Mean

We have now verified that the minimal triangulation lengths σ and σ_B are smooth subadditive and superadditive Euclidean functionals of order 1, respectively. We conclude the proof of Theorem 9.1 by showing closeness in mean (3.15) of $\sigma := \sigma_\delta$ and $\sigma_B := \sigma_{B,\delta}$. Letting $U_1, ..., U_n$ denote i.i.d. random variables with the uniform distribution on $[0,1]^2$, we will actually establish the stronger bound

$$(9.13) \qquad |E\sigma(U_1, ..., U_n) - E\sigma_B(U_1, ..., U_n)| \leq C(n \log n)^{1/4}$$

which will be useful in obtaining the rate (9.2).

Given $\varepsilon > 0$ and the random variables $U_1, ..., U_n$ we recall that $G_\varepsilon := G_\varepsilon(U_1, ..., U_n)$ are the uniquely defined Steiner sets with the property that

$$|S_B(\{U_1, ..., U_n\} \cup G_\varepsilon)| \leq \sigma_B(U_1, ..., U_n) + \varepsilon.$$

Let $\Delta_B^\varepsilon(U_1, ..., U_n)$ denote the collection of polygons generated by the boundary triangulations $S_B(\{U_1, ..., U_n\} \cup G_\varepsilon)$ and let $\mathcal{E}^\varepsilon(U_1, ..., U_n)$ be the collection of edges formed from the intersection of $\Delta_B^\varepsilon(U_1, ..., U_n)$ and the interior of $[0,1]^2$. To prove (9.13) we first bound the lengths of the edges in $\mathcal{E}^\varepsilon(U_1, ..., U_n)$. This edge length bound implies that with high probability the collection $\Delta_B^\varepsilon(U_1, ..., U_n)$ contains neither hexagons nor those pentagons with a side linking opposite sides of $[0,1]^2$. Thus with high probability there are at most four pentagons in $\Delta_B^\varepsilon(U_1, ..., U_n)$.

Lemma 9.8. *(edge length bounds) With high probability all edges $E \in \mathcal{E}^\varepsilon(U_1, ..., U_n)$ have a length $|E|$ satisfying*

$$(9.14) \qquad\qquad |E| \leq C(\log n/n)^{1/2}.$$

The meaning of the high probability statement (9.14) is: for any prescribed $\alpha > 0$ we can find $C := C(\alpha) > 0$ and a set Ω_0 with $P\{\Omega_0^c\} = O(n^{-\alpha})$ such that on Ω_0 all edges $E \in \mathcal{E}^\varepsilon(U_1, ..., U_n)$ satisfy the bound

$$|E| \leq C(\log n/n)^{1/2}.$$

The following argument can be modified to show this precise statement.

Proof. (sketch) The proof is a simple consequence of the fact that the aspect ratios of the polygons in $\Delta_B^\varepsilon(U_1, ..., U_n)$ are bounded and therefore if an edge E belongs to $\mathcal{E}^\varepsilon(U_1, ..., U_n)$, then there is a ball of radius $C|E|$ which is contained in $[0,1]^2$ and which does not contain any sample points, where $C := C(\delta)$ is a constant depending only on δ.

Indeed, for all $x \in [0,1]^2$ and $r > 0$, let $B(x, r)$ designate the ball centered at x with radius r and let $E_n(r)$ denote the event that there is an edge $E \in \mathcal{E}^\varepsilon(U_1, ..., U_n)$ whose length exceeds r. Given $E_n(r)$, the bounded aspect ratio assumption implies the existence of a ball of radius at least Cr which is contained entirely within a polygon and thus does not contain any sample points. This is clearly true for edges E which do not meet the boundary. For edges E meeting the boundary there are several cases which may be checked in a straightforward fashion. In any case, there is a $C < \infty$ such that

$$(9.15) \qquad E_n(r) \subset \{\exists\, x \in [0,1]^2 : B(x, Cr) \subset [0,1]^2,\ B(x, Cr) \cap \{U_i\}_{i \leq n} = \emptyset\}.$$

Thus, $P\{E_n(r)\}$ is bounded by the probability that there is "hole" of radius at least Cr in the sample $\{U_1, ..., U_n\}$. It is well-known and easy to show that with high probability holes with radius larger than $C(\log n/n)^{1/2}$ do not exist. Thus, with high probability, edges in $\mathcal{E}^\varepsilon(U_1, ..., U_n)$ have length less than $C(\log n/n)^{1/2}$. $\square$

We require one more auxiliary result before proving (9.13). To simplify the notation, write $\sigma(n)$ for $\sigma(U_1, .., U_n)$ and likewise for $\sigma_B(n)$ and $\mathcal{E}^\varepsilon(n)$. Let $S_B^\varepsilon(n) := S_B(\{U_1, ..., U_n\} \cup G_\varepsilon)$. Consider the edges in $\mathcal{E}^\varepsilon(n)$ which meet the boundary of $[0,1]^2$ and let $\Sigma^\varepsilon(n) := \Sigma^\varepsilon(U_1, ..., U_n)$ denote the sum of the lengths of these edges. The following lemma gives a crude yet sufficient upper bound for $\Sigma^\varepsilon(n)$.

Lemma 9.9. *For all $0 < \varepsilon < 1$, $E\Sigma^\varepsilon(n) \leq C(n \log n)^{1/4}$.*

Proof. Decompose $[0,1]^2$ into a subsquare R_1 and a moat $R_2 := [0,1]^2 - R_1$; choose R_1 so that it has side length $1 - C(\log n/n)^{1/2}$ and is centered within $[0,1]^2$. Let $|S_B^\varepsilon(n) \cap R_1|$ denote the sum of the lengths of the edges in $S_B^\varepsilon(n) \cap R_1$ and similarly for $|S_B^\varepsilon(n) \cap R_2|$. By Lemma 9.8 we have that $\Sigma^\varepsilon(n) \leq |S_B^\varepsilon(n) \cap R_2|$ with high probability. It will thus be enough to show

$$E|S_B^\varepsilon(n) \cap R_2| \leq C(n \log n)^{1/4}.$$

Since $S_B^\varepsilon(n) \cap R_1$ is a feasible boundary triangulation of $\{U_1, ..., U_n\} \cap R_1$ with respect to R_1 we have

$$|S_B^\varepsilon(n)| = |S_B^\varepsilon(n) \cap R_1| + |S_B^\varepsilon(n) \cap R_2|$$
$$\geq \sigma_B(\{U_1, ..., U_n\} \cap R_1, R_1) + |S_B^\varepsilon(n) \cap R_2|$$

where the inequality follows by the minimality of σ_B.

The number of sample points in R_1 is a binomial random variable $B(n, p)$ with parameters n and p, $p :=$ area R_1. Taking expectations and scaling we get

$$E|S_B^\varepsilon(n)| \geq \left(1 - C(\log n / n)^{1/2}\right) E\sigma_B(\{U_1, ..., U_{B(n,p)}\}) + E|S_B^\varepsilon(n) \cap R_2|.$$

By definition we have $E|S_B^\varepsilon(n)| \leq E\sigma_B(n) + \varepsilon$ and thus

$$E|S_B^\varepsilon(n) \cap R_2| \leq E\sigma_B(n) + \varepsilon - \left(1 - C(\log n / n)^{1/2}\right) E\sigma_B(B(n, p))$$
$$\leq E\left(\sigma_B(n) - \sigma_B(B(n, p))\right) + \varepsilon + C(\log n / n)^{1/2} E(B(n, p)^{1/2})$$

by the growth bound (9.6). By the smoothness of σ_B, the above is bounded by

$$\leq CE(|n - B(n, p)|^{1/2}) + \varepsilon + C(\log n / n)^{1/2}(np)^{1/2}$$
$$\leq C\left(E|B(n, 1 - p)|\right)^{1/2} + C(\log n)^{1/2}$$
$$\leq C\left(n(1 - p)\right)^{1/2} + C(\log n)^{1/2}.$$

Since $p := (1 - C(\log n / n)^{1/2})^2 \geq 1 - C(\log n / n)^{1/2}$, we see that $n(1 - p) \leq C(n \log n)^{1/2}$, completing the proof of Lemma 9.9. $\square$

We are now positioned to establish the estimate (9.13) and thus conclude the proof of Theorem 9.1.

Lemma 9.10. (*closeness in mean*) We have

$$|E\sigma(U_1, ..., U_n) - E\sigma_B(U_1, ..., U_n)| \leq C(n \log n)^{1/4}.$$

Proof. It suffices to show for all $0 < \varepsilon < 1$ that

$$\sigma_B(n) \leq \sigma(n) \leq \sigma_B(n) + \Sigma^\varepsilon(n) + C.$$

Lemma 9.10 follows since it is enough to take expectations and apply Lemma 9.9. Let $0 < \varepsilon < 1$ and $S_B^\varepsilon(n)$ be as above. We claim that

$$\sigma_B(n) \leq \sigma(n) \leq |S_B^\varepsilon(n)| + \Sigma^\varepsilon(n) + C.$$

The first inequality follows by the definition of σ_B. To prove the second, we need to show that there is a feasible triangulation of $\{U_1, ..., U_n\}$ whose total length is bounded by $|S_B^\varepsilon(n)| + \Sigma^\varepsilon(n) + C$. Such a triangulation is obtained by triangulating the quadrilaterals and pentagons in the graph described by $S_B^\varepsilon(n)$ (recall that with high probability the graph contains no hexagons and at most four pentagons). We triangulate the quadrilaterals by adding their diagonals, the sum of whose lengths is at most the sum of $\Sigma^\varepsilon(n)$ and the perimeter of the unit square. A triangulation of the pentagons may be achieved with a cost bounded by a constant since there are at most four pentagons. Thus we have shown the claim. Since $|S_B^\varepsilon(n)| \leq \sigma_B(n) + \varepsilon$, the result follows. $\square$

We have now proved Theorem 9.1 and turn to the proof of Theorem 9.2. The proof depends on the following general rate result, which follows from a slight modification of the proof of Theorem 5.2.

Theorem 9.11. *(rates of convergence) Let L be a smooth subadditive Euclidean functional of order 1 on $[0,1]^2$ such that the following "add-one bound" is satisfied:*

$$(9.16) \qquad |EL(U_1, ..., U_n) - EL(U_1, ..., U_{n+1})| \leq C(\log n/n)^{1/2}.$$

If $|EL(U_1, ..., U_n) - EL_B(U_1, ..., U_n)| \leq \beta(n)$ where $\beta(n)$ is a function of n, then

$$(9.17) \qquad |EL(U_1, ..., U_n) - \alpha(L)n^{1/2}| \leq C\left(\beta(n) \vee (\log n)^{1/2}\right).$$

Proof of Theorem 9.2. To apply Theorem 9.11 to triangulations, we must show that σ satisfies the estimate (9.16). For all $\varepsilon > 0$ and $U_1, ..., U_n$ we recall that $G_\varepsilon := G_\varepsilon(U_1, ..., U_n)$ are (uniquely defined) Steiner sets with the property that

$$|S(\{U_1, ..., U_n\} \cup G_\varepsilon)| \leq \sigma(U_1, ..., U_n) + \varepsilon.$$

Recall that $\Delta^\varepsilon(U_1, ..., U_n)$ denotes the collection of triangles generated by $S(\{U_1, ..., U_n\} \cup G_\varepsilon)$. Notice that

$$\begin{aligned}
\sigma(U_1, ..., U_n) &\leq \sigma(U_1, ..., U_{n+1}) \\
&\leq |S(\{U_1, ..., U_n\} \cup G_\varepsilon)| + 3D(\varepsilon, n) \\
&\leq \sigma(U_1, ..., U_n) + 3D(\varepsilon, n) + \varepsilon,
\end{aligned}$$
(9.18)

where $D(\varepsilon, n)$ is the diameter of the *random* triangle in $\Delta^\varepsilon(U_1, ..., U_n)$ which contains the point U_{n+1}. By Lemma 9.8 we have $D(\varepsilon, n) \leq C(\log n/n)^{1/2}$ with high probability where C doesn't depend upon ε. Letting $\varepsilon = (\log n/n)^{1/2}$ gives the high probability bound

$$|\sigma(U_1, ..., U_n) - \sigma(U_1, ..., U_{n+1})| \leq C(\log n/n)^{1/2}$$

and therefore the add-one bound

$$|E\sigma(U_1, ..., U_n) - E\sigma(U_1, ..., U_{n+1})| \leq C(\log n/n)^{1/2}.$$

Thus, by (9.13) and (9.17) we obtain

$$|E\sigma(U_1, ..., U_n) - \alpha(\sigma)n^{1/2}| \leq C(n \log n)^{1/4}$$

which is the desired estimate (9.2). This concludes the proof of Theorem 9.2. $\square$

9.6. The Probabilistic Plateau Functional

To establish Theorem 9.3 we may adapt the above approach to the three dimensional setting. We fix D once and for all, $1 < D < \infty$. It is first helpful to enlarge the definition of $\tau := \tau_D$ in the following way. Let $F \subset \Delta$ be a finite set, where $\Delta \subset [0, 1]^3$ is a convex polyhedron. Given D, a tetrahedralization $T_D(F, \Delta)$ is a decomposition of Δ into tetrahedra with aspect ratios bounded by D such that the tetrahedral vertices coincide with F and the corners of Δ. Analogously to $\tau_D(F)$, we define

$$\tau_D(F, \Delta) := \inf_G |T_D(F \cup G, \Delta)|,$$

where $|T_D(F \cup G, \Delta)|$ denotes the total surface area and where G ranges over finite sets in Δ.

We now consider the properties of the Plateau functional $\tau := \tau_D$. Notice that $\tau(\alpha F) = \alpha^2 \tau(F)$ and thus τ is homogeneous of order 2. Modifications of the proof of Lemma 9.6 show that τ is subadditive on $\mathbb{R}^3$. Also, for any $F \subset [0, 1]^3, F \neq \emptyset$, we have $\tau(F) \leq C(\text{card} F)^{1/3}$ by Lemma 3.3 (here and in all that follows, $C := C(D)$ denotes a constant depending only on D and whose value may vary from line to line).

To see that τ is a smooth subadditive Euclidean functional on $\mathbb{R}^3$ of order two, it remains only to verify smoothness

$$|\tau(F \cup G) - \tau(F)| \leq C(\text{card} G)^{1/3}.$$

We will closely follow the approach used to verify the smoothness of σ. For all $\varepsilon > 0$ and $F \subset [0, 1]^3$ let $G_\varepsilon := G_\varepsilon(F)$ be Steiner sets such that

$$|T_D(F \cup G_\varepsilon)| \leq \tau_D(F) + \varepsilon.$$

Let $T^\varepsilon(F) := T_D(F \cup G_\varepsilon)$. Let $\Delta^\varepsilon(F)$ denote the collection of tetrahedra defined by $T^\varepsilon(F)$ and let $\mathcal{E}^\varepsilon := \mathcal{E}^\varepsilon(F)$ denote the collection of tetrahedral edges. Since the tetrahedra have uniformly bounded aspect ratios, the sum of the cubes of the edges in $\mathcal{E}^\varepsilon$ is bounded by a finite constant C, $C := C(D)$. Using the growth bound $\tau(F) \leq C(\text{card} F)^{1/3}$, we easily establish the analog of (9.11), namely

$$\tau(F \cup G) \leq |T^\varepsilon(F)| + C \sum_{i=1}^{J} |E_i|^2 (\text{card}(G \cap \Delta_i))^{1/3},$$

where E_i is the longest edge of tetrahedron Δ_i, $\Delta_i \in \Delta^\varepsilon(F)$. Hölder's inequality, together with the bound $\sum_{i=1}^{J} |E_i|^3 \leq C$, completes the proof of smoothness.

Given a convex polyhedron $\Delta \subseteq [0,1]^3$ and $F \subset \mathbb{R}^3$ such that $\mathrm{co}(F) \supset \Delta$, we next define a "boundary tetrahedralization" of Δ with respect to F in the same way that we defined boundary triangulations. Given $1 < D < \infty$ fixed, consider a D-tetrahedralization of $\mathrm{co}(F)$. Such a tetrahedralization partitions $\mathrm{co}(F)$ into tetrahedra with aspect ratios bounded by D and whose vertices coincide with F. The boundary tetrahedralization $T_{B,D}(F,\Delta)$ of Δ with respect to F is obtained by considering the intersection of Δ and the D-tetrahedralization of $\mathrm{co}(F)$. The boundary tetrahedralization of Δ thus generates the usual tetrahedra as well as polyhedra with faces contained in the boundary of Δ. We let $|T_{B,D}(F)|$ denote the total surface area of the faces of the polyhedra in the interior of Δ. We let

$$\tau_{B,D}(F,\Delta) := \inf_G |T_{B,D}(F \cup G, \Delta)|,$$

where G ranges over all finite sets in $\mathbb{R}^3$ with the property that $\mathrm{co}(F \cup G) \supset \Delta$. We will suppress mention of D and henceforth write $\tau_B(F)$ for $\tau_{B,D}(F, [0,1]^3)$ and $\tau_B(F,\Delta)$ instead of $\tau_{B,D}(F,\Delta)$.

Given this extended definition of τ_B, observe that τ_B is a smooth superadditive Euclidean functional on $\mathbb{R}^3$ of order two. To verify smoothness

$$|\tau_B(F \cup G, [0,1]^3) - \tau_B(F, [0,1]^3)| \leq C(\mathrm{card}\,G)^{1/3}$$

we may follow the approach of section four.

We may also show closeness in mean (3.15) of the functionals τ and τ_B with $p = 2$ and $d = 3$ there. To do this, we will follow the approach of section five. For all $\varepsilon > 0$ and $U_1, ..., U_n$, we let $G_\varepsilon(U_1, ..., U_n)$ be uniquely defined Steiner sets such that

$$(9.19) \qquad |T_B(\{U_1, ..., U_n\} \cup G_\varepsilon)| \leq \tau_B(U_1, ..., U_n) + \varepsilon.$$

Let $T_B^\varepsilon(n) := T_B(\{U_1, ..., U_n\} \cup G_\varepsilon)$. Let Δ_B^ε denote the collection of polyhedra generated by the tetrahedralization $T_B^\varepsilon(n)$. Let $\mathcal{E}^\varepsilon := \mathcal{E}^\varepsilon(U_1, ..., U_n)$ be the collection of edges formed from the intersection of the edges in Δ_B^ε and the interior of $[0,1]^3$ and let $\mathcal{F}^\varepsilon := \mathcal{F}^\varepsilon(U_1, ..., U_n)$ be the collection of faces formed by intersecting the faces of Δ_B^ε and the interior of $[0,1]^3$. The following estimate is the analog of Lemma 9.8. Observe that the proof of (9.20) below follows the proof of (9.14) with small modifications. Note that (9.21) follows from (9.20) and the bounded aspect ratio property of the tetrahedra.

Lemma 9.12. *With high probability all edges $E \in \mathcal{E}^\varepsilon$ satisfy the length bound*

$$(9.20) \qquad |E| \leq C(\log n/n)^{1/3}$$

and all faces $F \in \mathcal{F}^\varepsilon$ satisfy the area bound

$$(9.21) \qquad \mathrm{area}\,F \leq C(\log n/n)^{2/3}.$$

Consider the faces in $\mathcal{F}^\varepsilon$ which meet the boundary of $[0,1]^3$ and let $\Sigma^\varepsilon(n) := \Sigma^\varepsilon(U_1, ..., U_n)$ denote the sum of their areas. Exactly as in Lemma 9.9, we may find a rough estimate for $\Sigma^\varepsilon(n)$.

Lemma 9.13. *For all $0 < \varepsilon < 1$ we have $E\Sigma^\varepsilon(n) \leq Cn^{2/9}(\log n)^{1/9}$.*

Proof. We will follow the proof of Lemma 9.9. Decompose $[0,1]^3$ into a subcube Q_1 centered within $[0,1]^3$ and a moat $Q_2 := [0,1]^3 - Q_1$. Let the edge length of Q_1 be $1 - C(\log n/n)^{1/3}$. Let $|T_B^\varepsilon(n) \cap Q_1|$ denote the sum of the areas of the faces in $T_B^\varepsilon(n) \cap Q_1$ and similarly for $|T_B^\varepsilon(n) \cap Q_2|$. By Lemma 9.13 we have $\Sigma^\varepsilon(n) \leq |T_B^\varepsilon(n) \cap Q_2|$ with high probability. It will be enough to show that

$$E|T_B^\varepsilon(n) \cap Q_2| \leq Cn^{2/9}(\log n)^{1/9}.$$

Now as in the proof of Lemma 9.9 we have

$$|T_B^\varepsilon(n)| \geq \tau_B(\{U_1, ..., U_n\} \cap Q_1, Q_1) + |T_B^\varepsilon(n) \cap Q_2|$$

and taking expectations we get

$$E|T_B^\varepsilon(n)| \geq (1 - C(\log n/n)^{1/3})^2 E\tau_B(\{U_1, ..., U_{B(n,p)}\}) + E|T_B^\varepsilon(n) \cap Q_2|,$$

where $B(n,p)$ denotes a binomial random variable with parameters n and $p :=$ volume of $Q_1 := (1 - C(\log n/n)^{1/3})^3$. Thus by scaling and the growth bounds for τ_B we have

$$\begin{aligned}
E|T_B^\varepsilon(n) \cap Q_2| &\leq E\tau_B(n) + \varepsilon - (1 - C(\log n/n)^{1/3})^2 E\tau_B(B(n,p)) \\
&\leq C(E|n - B(n,p)|^{1/3}) + \varepsilon + C(\log n/n)^{1/3}(np)^{1/3} \\
&\leq C(n(1-p))^{1/3} + C(\log n)^{1/3} \\
&\leq (nC(\log n/n)^{1/3})^{1/3} + C(\log n)^{1/3} \\
&\leq Cn^{2/9}(\log n)^{1/9}.
\end{aligned}$$

This completes the proof of Lemma 9.13 □

Let $\tau(n) := \tau(U_1, ..., U_n)$. The analog of Lemma 9.10 becomes

Lemma 9.14. *(closeness in mean) We have*

$$|E\tau(U_1, ..., U_n) - E\tau_B(U_1, ..., U_n)| \leq Cn^{2/9}(\log n)^{1/9}.$$

Proof. It suffices to show for all $0 < \varepsilon < 1$ that

$$\tau_B(n) \leq \tau(n) \leq \tau_B(n) + C(\Sigma^\varepsilon(n) + 1).$$

Lemma 9.14 follows by taking expectations and applying Lemma 9.13.

Let $0 < \varepsilon < 1$ and $T_B^\varepsilon(n)$ be as in (9.19). We claim that

(9.22) $$\tau_B(n) \leq \tau(n) \leq |T_B^\varepsilon(n)| + C(\Sigma^\varepsilon(n) + 1).$$

The first inequality follows by definition. To prove the second, we need to show that there is a feasible tetrahedralization of $\{U_1, ..., U_n\}$ whose total surface area is bounded by $|T_B^\varepsilon(n)| + C(\Sigma^\varepsilon(n) + 1)$. We observe that the polyhedra given by $T_B^\varepsilon(n)$ which meet the boundary are convex and may be tetrahedralized at a cost bounded by a constant multiple of the sum of the area of their faces. The combined areas of their faces is the sum of $\Sigma^\varepsilon(n)$ and the area of the boundary of $[0,1]^3$. This proves (9.22). Combining (9.22) and (9.19) we obtain Lemma 9.14. □

Lemma 9.14 establishes that the Plateau functional τ and its boundary version τ_B are close in mean (3.15) when $p = 2$ and $d = 3$. We have therefore shown that τ and τ_B satisfy all the conditions of Theorem 7.1 with $p = 2$ and $d = 3$. Theorem 9.3 follows.

It is a simple matter to find rates of convergence for the mean of τ. Since we are in dimension three we can avoid appealing to Theorem 9.11. We may use the subadditivity of τ, the superadditivity of τ_B, and Lemma 9.14 to obtain the rate estimate

$$|E\tau(U_1, ..., U_n) - \alpha(\tau)n^{1/3}| \le Cn^{2/9}(\log n)^{1/9}.$$

It is not clear whether these rates are optimal.

Notes and References

1. This chapter is based heavily on Yukich (1997a). Theorem 9.1 adds to Steele (1982) who considers the case $\delta = \infty$, i.e. the case involving no restrictions on the aspect ratios of the triangles. Steele uses geometric subadditivity of triangulations to establish (9.1) for σ_∞ in the special case that $X_1, ..., X_n$ are uniformly distributed on $[0,1]^2$. According to Steele (1982), Theorem 9.1 addresses a problem of György Turán.

2. Using the theory of smooth subadditive and superadditive Euclidean functionals, we have provided the asymptotics for the Steiner triangulation functional as well as its three-dimensional counterpart, the discrete Plateau functional. With regard to the former, we have extended the work of Steele (1982) under the regularity condition that the triangles have bounded aspect ratios. Whether one can remove or relax this condition remains open. Additional questions which merit investigation include:

(i) is there a suitable analog of Theorems 9.1 and 9.3 for triangulation and tetrahedralization functionals which do not use Steiner points?

(ii) are there asymptotics for triangulation and tetrahedralization functionals which are defined in terms of simplicial complexes?

10. GEOMETRIC LOCATION PROBLEMS

10.1. Statement of Problem

The purpose of this chapter is to develop the asymptotics of geometric location problems on a random sample. We use the general structure developed in the earlier chapters to obtain the asymptotics for the length of the graphs defined by geometric location problems.

Given n points in $\mathbb{R}^d$, $d \geq 2$, geometric location problems essentially involve choosing a subset of size k which "best represents the set". Such problems, also termed "k-median problems", are among the oldest in combinatorial optimization and have been studied by Fermat, Steiner, and Steinhaus (1956) among many others.

Let us now define our terms. Given a set $F := \{x_1, ..., x_n\}$ of points in $\mathbb{R}^d$ and $k \in \mathbb{N}$, choose k points in F as "medians" or "centers" and join the remaining vertices to the nearest center. The cost of serving a vertex equals the Euclidean distance to the nearest center. Letting $\mathcal{C}$ denote the collection of medians, the total cost of serving the points in F is thus

$$\sum_{i=1}^{n} \min_{x_j \in \mathcal{C}} \|x_i - x_j\|.$$

The k-median problem involves choosing the set $\mathcal{C}$ of medians so as to minimize the above sum. The *minimal cost* obtained by optimally choosing $\mathcal{C}$ is given by the k median functional

$$(10.1) \qquad M(k; F) := \inf_{\mathcal{C} \in \mathbb{C}} \sum_{i=1}^{n} \min_{x_j \in \mathcal{C}} \|x_i - x_j\|,$$

where $\mathbb{C} := \mathbb{C}(k)$ denotes the collection of all subsets $\mathcal{C}$ of F of cardinality k. There is no restriction on the number of sites served by each center. See Figure 10.1.

When F consists of a three point set in $\mathbb{R}^2$, $k = 1$, and $\mathbb{C} := \mathbb{C}(1)$ ranges over all singletons which need not belong to F, then (10.1) reduces to a well-known problem of Fermat: "given three points in the plane, find a fourth point such that the sum of its distances to the three given points is a minimum". This problem has been widely popularized by Courant and Robbins (1941) and we refer to Kuhn (1974) for more on its history and applications.

The k-median problem is computationally difficult and Papadimitriou (1981) showed that it is NP-complete, resolving a conjecture of Fisher and Hochbaum (1980).

Figure 10.1. A k-median graph on 12 points with k = 3

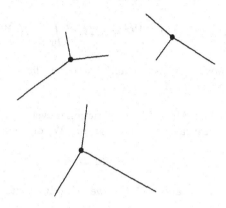

In this chapter we describe the asymptotic behavior of a version of the k-median functional on random points in the unit cube. We claim that a modification of the k-median problem fits neatly into the theory of subadditive and superadditive Euclidean functionals and in this way we describe its stochastic behavior. This adds to the work of Hochbaum and Steele (1982), who made the first progress in the study of the stochastic analysis of the k-median problem.

Hochbaum and Steele (1982) showed that if $U_1, ..., U_n$ are i.i.d. random variables with the uniform distribution on $[0, 1]^2$ then the functional $M(k; U_1, ..., U_n)$ behaves like a smooth subadditive Euclidean functional of order 1 when k grows linearly with n. More precisely they showed:

Theorem 10.1. *(Hochbaum and Steele, 1982) If $0 < \alpha < 1$ then*

(10.2) $$\lim_{n \to \infty} M(\alpha n; U_1, ..., U_n)/n^{1/2} = C \quad a.s.$$

where $C := C(\alpha)$ is a finite non-zero constant.

We now consider a natural modification of the k-median problem. Let $|F| :=$ card$F$. Given $D \geq 2$, and F a finite set in $\mathbb{R}^d$, $d \geq 2$, consider the functional $M(\lceil |F|/(D + 1)\rceil; F)$ under the additional assumption that no center serves more than D sites in F in addition to itself. Here $\lceil x \rceil$ denotes the smallest integer greater than or equal to x. Thus the degree of each center is at most D. Were we to consider $M(\lfloor |F|/(D+1)\rfloor; F)$ then there may not be enough centers of degree D to serve the points in F. Let $M(D; F)$ henceforth denote $M(\lceil |F|/(D + 1)\rceil; F)$.

We claim that $M(D; F)$ is a subadditive Euclidean functional and has a canonical boundary functional $M_B(D; F)$ which is superadditive, smooth, and close in mean (3.15) to $M(D; F)$. For details we refer to McGivney and Yukich (1997). In this way we obtain rate results for $M(D; F)$ (Chapter 5), large deviations (Chapter 6), and the following asymptotics (Chapter 7):

Theorem 10.2. (*asymptotics for geometric location problems*) *Let* X_i, $i \geq 1$, *be i.i.d. random variables with values in* $[0,1]^d$, $d \geq 2$. *Then*

$$\lim_{n \to \infty} M(D; X_1, ..., X_n)/n^{(d-1)/d} = \alpha(M, d) \int_{[0,1]^d} f(x)^{(d-1)/d} dx \quad c.c.,$$

where $\alpha(M, d)$ *is a positive constant and* f *denotes the density of the absolutely continuous part of the law of* X_1.

It is convenient to view $M(D; F)$ as a functional over the rectangles $\mathcal{R}(d)$ in the way that has become natural in this monograph. We define for all $R \in \mathcal{R}(d)$

(10.3) $$M(D; F, R) := M(D; F \cap R).$$

When R is the unit cube $[0,1]^d$ we simply write $M(D; F)$ instead of $M(D; F, [0,1]^d)$.

The proof of Theorem 10.2 depends heavily on introducing a *boundary functional* $M_B(D; F, R)$ associated with $M(D; F, R)$. Boundary functionals are a key conceptual and technical tool in the study of optimization problems and this is the case here as well. There is more than one way to define a boundary k-median functional. It turns out that the following definition insures superadditivity (3.3) and closeness in mean (3.15).

For all rectangles $R \in \mathcal{R} := \mathcal{R}(d)$ let R' denote the enlarged rectangle $\{x \in \mathbb{R}^d : d(x, R) \leq 1\}$, where $d(x, R)$ denotes the Euclidean distance between x and R. For all finite subsets S of the moat $R' - R$ consider the graph G which realizes $M(D; F \cup S, R')$ and let $G_R := G \cap R$ be its restriction to R. Denote the length of G_R by $M_R(D; F \cup S, R')$. Since the points in S are arbitrary, G_R may have overlapping edges and there may be as many as D copies of an edge joining a point in F to a point on ∂R. Let

(10.4) $$M_B(D; F, R) := \inf_S M_R(D; F \cup S, R'),$$

where the infimum ranges over all finite subsets S of $R' - R$ having rational coordinates. This last requirement ensures the measurability of $M_B(D; X_1, ..., X_n, R)$ where X_i, $i \geq 1$, are random variables.

The bounded degree requirement on the centers implies that the subsets S of $R' - R$ have cardinality at most $D|F|$. The infimum (10.4) is thus realized by some set S. We call M_B the *boundary k-median functional*. Notice that $M_B \leq M$. As indicated, the boundary functional M_B is smooth of order 1, smooth, superadditive, and close in mean to the standard functional M; see McGivney and Yukich (1997b).

10.2.   Additional Remarks

Rates of Convergence. It is natural to look for a rate of convergence in Theorem 10.2. To find a rate of convergence in the uniform case we apply the following lemma, essentially a modification of Theorem 5.1 when $p = 1$. This lemma is useful whenever the bound for

$$|EL(U_1, ..., U_n) - EL_B(U_1, ..., U_n)|$$

is different from $Cn^{(d-2)/d}$.

Lemma 10.3. *(rates of convergence of means) Let $U_1, ..., U_n$ be i.i.d. uniform random variables on $[0,1]^d$, $d \geq 3$. Suppose that L is a smooth subadditive Euclidean functional, L_B is a smooth superadditive Euclidean functional and*

$$(10.5) \qquad |EL(U_1, ..., U_n) - EL_B(U_1, ..., U_n)| \leq \beta(n),$$

where $\beta(n)$ denotes a function of n. Then

$$|EL(U_1, ..., U_n) - \alpha(L, d)n^{(d-1)/d}| \leq \beta(n) \vee Cn^{(d-1)/2d}.$$

As shown in McGivney and Yukich (1997b), the functionals M and M_B satisfy (10.5) with $\beta(n) := Cn^{((d-1)/d)^2}$. We immediately obtain the rate result

$$|EM(D; U_1, ..., U_n) - \alpha(M, d)n^{(d-1)/d}| \leq Cn^{((d-1)/d)^2}.$$

It is not clear whether this rate estimate can be improved to give an error as small as $O(n^{(d-2)/d})$, which would be consistent with the rate results for the TSP and MST obtained in Chapter 5.

Generalizations. The methods above apply to at least one natural modification of the present problem. Consider the *Steiner k-median problem* which is a generalization of Fermat's problem and the bounded degree k-median problem. It is defined in the following way. Given points F in $[0,1]^d$, choose $k := \lceil \frac{|F|}{D+1} \rceil$ centers either from F or from Steiner points in $[0,1]^d$. Join each of the non-center points in F to its closest center under the restriction that no more than $D + 1$ points can be joined to any center. A component is said to be complete if the degree of the center is $D + 1$ and incomplete otherwise. If a center comes from F then its degree equals the number of vertices it serves, *including* itself. This will ensure that the number of incomplete components is at most D.

Define the *Steiner k-median functional* by

$$M^s(D; F, R) := \inf_S M(d; F \cup S, R),$$

where the infimum runs over all sets S of Steiner points in R. As in the definition of $M_B(D; F, R)$, observe that the infimum is realized by some set S.

Next we define the *boundary Steiner k-median functional* $M^s_B(D; F, R)$. For all rectangles $R \in \mathcal{R}$, let R' denote the enlarged rectangle $\{x \in \mathbb{R}^d : d(x, R) \leq 1\}$. For all finite sets S' of points belonging to the moat $R' - R$, consider the graph G which realizes $M^s(D; F \cup S', R')$ and let its restriction to R have length $M^s_R(D; F \cup S', R')$. Since the points in S are arbitrary, the restriction of G to R may produce as many as D copies of an edge which joins a point in F to ∂R. Let

$$M^s_B(D; F, R) := \inf_{S'} M(D; F \cup S, R),$$

where the infimum runs over all finite subsets S' of $R' - R$ having rational co-ordinates. We call M^s_B the *boundary Steiner k-median functional*. Notice that $M^s_B \leq M^s$.

Modifications of the proofs of simple subadditivity, subadditivity, and smoothness for $M(D; F, R)$ (see McGivney and Yukich (1997b)) show that $M^s(D; F, R)$ is a smooth subadditive Euclidean functional. We will sketch the proof of geometric subadditivity (3.4). For each $1 \leq i \leq 2$ consider the local k-median graphs G_i given by $M^s(D; F \cap R_i, R_i)$. There are at most D^2 *non-Steiner* vertices which are elements of incomplete components in each such graph. Therefore there are at most $2D^2$ *non-Steiner* vertices $\mathcal{V}$ which belong to incomplete components in $G_1 \cup G_2$. Construct a k-median graph $G_\mathcal{V}$ on $\mathcal{V}$. This graph together with the union of the complete components in $G_1 \cup G_2$ gives a feasible k-median graph. Since the length of $G_\mathcal{V}$ is bounded by a constant we have established subadditivity (3.4).

Likewise, minor changes in the proofs of superadditivity and smoothness for $M_B(D; F, R)$ show that $M^s_B(D; F, F)$ is also a smooth superadditive Euclidean functional. Since $M^s(D; F, R)$ and $M^s_B(D; F, R)$ are also close in mean, the umbrella Theorem 7.1 implies that

$$\lim_{n \to \infty} M^s(D; X_1, ..., X_n)/n^{(d-1)/d} = \alpha(M^s, d) \int_{[0,1]^d} f(x)^{(d-1)/d} dx \quad c.c.,$$

where $\alpha(M^s, d)$ is a positive constant and f denotes the density of the absolutely continuous part of the law of X_1. $\quad\square$

Notes and References

1. There are several possibilities for extending the results of this chapter. Proving Theorem 10.2 without any assumption on the number of sites served by a center would be worthwhile. We do not address the case of power-weighted edges. Such an extension seems straightforward.

2. Most of this chapter is based on McGivney and Yukich (1997b).

11. WORST CASE GROWTH RATES

11.1. Introduction

Previous chapters focused on the behavior of the *stochastic* versions of problems in Euclidean combinatorial optimization. The notion of boundary functionals has been central to our analysis. Boundary functionals, with their intrinsic superadditivity, are the key to proving general umbrella theorems for power-weighted versions of Euclidean functionals. They are also the basis for finding rates of convergence and for developing ergodic theorems.

In this chapter we will see that boundary functionals are valuable tools in a setting that involves no probabilistic assumptions. Boundary functionals turn out to be valuable in describing growth rates of the worst case values of many of the classic problems in combinatorial optimization and operations research. It is somewhat surprising that the asymptotics for the worst case lengths match those found in the probabilistic setting.

Throughout we focus on the following *worst case* counterparts of the TSP, MST, and minimal matching functionals. We let $|V|$ denote the cardinality of the set V.

Definition 11.1.

(i) The largest possible length of any minimal traveling salesman tour (worst case tour) with pth power weighted edges formed from n points in $[0,1]^d$ is

$$(11.1) \qquad \tau^p(n) := \max_{V \subset [0,1]^d, |V|=n} T^p(V).$$

(ii) The largest possible length of any minimum spanning tree with pth power weighted edges formed from n points in $[0,1]^d$ is

$$(11.2) \qquad \mu^p(n) := \max_{V \subset [0,1]^d, |V|=n} M^p(V).$$

(iii) The largest possible length of any minimal matching with pth power weighted edges formed from n points in $[0,1]^d$ is

$$(11.3) \qquad \sigma^p(n) := \max_{V \subset [0,1]^d, |V|=n} S^p(V).$$

The functions $\tau^p(n)$, $\mu^p(n)$, and $\sigma^p(n)$ represent the worst case values of the TSP, MST, and minimal matching functionals, respectively. The worst case values of the boundary TSP, MST, and minimal matching functionals are defined similarly and are denoted by $\tau_B^p(n)$, $\mu_B^p(n)$, and $\sigma_B^p(n)$, respectively.

The points in V need not be distinct. Thus the worst case functions (11.1), (11.2) and (11.3) are continuous functions on the compact set formed from the product of n copies of $[0,1]^d$. These functions therefore exist and are monotone increasing, that is $\tau^p(n) \leq \tau^p(n+1)$ and similarly for μ^p and σ^p.

The bulk of this chapter provides an approach which yields asymptotics for the worst case functions (11.1) - (11.3). This approach also provides the asymptotics for the worst case versions of other Euclidean functionals, including the rectilinear Steiner tree, treated by Snyder (1992). We will not pursue this, but limit the discussion to the worst case versions (11.1) - (11.3) of the archetypical functions.

Our main result gives growth rates for the functions (11.1) - (11.3). These growth rates are identical to those in the basic limit Theorems 4.1 and 4.3 and the umbrella Theorem 7.1.

Theorem 11.2. Let $1 \leq p < d$. Let $\rho^p(n)$ denote either of the three worst case functions given by Definition 11.1. Then

$$(11.4) \qquad \lim_{n \to \infty} \rho^p(n)/n^{(d-p)/d} = \beta(\rho^p, d),$$

where $\beta(\rho^p, d)$ is a positive constant depending only on ρ^p and d.

Although worst case functions have received considerable attention, only Snyder (1987) and Steele and Snyder (1989) treat their asymptotics. We will not try to determine the values of the constant $\beta(\rho^p, d)$ for the different choices of p.

11.2. Superadditivity

The method for proving (11.4) uses a strategy that is by now familiar: use the intrinsic superadditivity of the worst case boundary functions ρ_B^p to easily establish that ρ_B^p satisfies the asymptotics (11.4) and then conclude the proof of (11.4) by checking that $\rho_B^p(n)$ is within $o(n^{(d-p)/d})$ of $\rho^p(n)$.

To see that (11.4) holds for the worst case boundary function ρ_B^p, we first show that ρ_B^p satisfies *monotonicity*, *boundedness*, and *superadditivity*. To check monotonicity, we have already seen that ρ is monotone and for the same reasons ρ_B^p is also monotone, i.e.,

$$(11.5) \qquad \rho_B^p(n) \leq \rho_B^p(n+1).$$

Recalling the growth bounds (3.7) of our archetypical functionals, we have

$$(11.6) \qquad \rho_B^p(n) \leq C_2 n^{(d-p)/d}.$$

We now check superadditivity, which says that for all $1 \leq p < d$ and positive integers n and k we have

(11.7) $$n^{d-p} \rho_B^p(k) \leq \rho_B^p(kn^d).$$

Let us check that (11.7) holds when ρ is the worst case MST function. We proceed as follows. Subdivide $[0,1]^d$ into n^d subcubes $Q_1, ..., Q_{n^d}$ of edge length $1/n$. In each subcube put k points in such a way that the worst case value of $n^{-p}\mu_B^p(k)$ is achieved. The union of these points over all n^d subcubes gives a set F of cardinality kn^d. By the definition of the worst case MST, we have clearly $M_B^p(F) \leq \mu_B^p(kn^d)$. It only remains to show that $M_B^p(F) \geq n^{d-p}\mu_B^p(k)$.

To see this last inequality, let G be the tree which realizes the boundary functional $M_B^p(F)$ on F. The restriction of G to each subcube Q_i, $1 \leq i \leq n^d$, generates a boundary rooted tree on $F \cap Q_i$. Every such tree has a length which is at least as large as $n^{-p}\mu_B(k)$, the length of the boundary rooted MST on $F \cap Q_i$. This holds for all $1 \leq i \leq n^d$, which shows that $n^{d-p}\mu_B^p(k) \leq M_B^p(F)$, as desired.

This simple argument shows that the worst case MST boundary function is super-additive. Modifications of this argument show that the worst case TSP and minimal matching boundary functionals are superadditive as well.

Conditions (11.5)-(11.7) are rather strong and it is not surprising that together they imply the existence of the limit

(11.8) $$\lim_{n\to\infty} \rho_B^p(n)/n^{(d-p)/d} = \beta(\rho_B^p, d).$$

To show (11.8), define for fixed $1 \leq p < d$ the function $\phi(n) := \rho_B^p(n)/n^{(d-p)/d}$. Set $\beta := \limsup_{n\to\infty} \phi(n)$ and note that $\beta < \infty$ by (11.6). We want to show that $\liminf_{n\to\infty} \phi(n) \geq \beta$.

To see this, note that condition (11.7) tells us that for all positive integers n and k we have

(11.9) $$\phi(n^d k) \geq \phi(k).$$

Now given $\epsilon > 0$, find $k_o := k_o(\epsilon)$ such that $\phi(k_o) \geq \beta - \epsilon$. Thus for all $n \geq 1$ and $k \geq k_o$, (11.9) gives

$$\phi(n^d k) \geq \beta - \epsilon.$$

We now use an interpolation argument and the assumed monotonicity (11.5) to deduce that $\phi(j) \geq \beta - 2\epsilon$ for all j sufficiently large. Indeed, find n_o such that $(\beta - \epsilon)(\frac{n}{n+1})^{d-p} \geq \beta - 2\epsilon$ holds for all $n \geq n_o$. Given $j \geq n_o^d k_o$, find the unique $n \geq n_o$ such that $n^d k_o \leq j < (n+1)^d k_o$. Then (11.5) implies

$$\phi(j) \geq \phi(n^d k_o)(\frac{n}{n+1})^{(d-p)} \geq \beta - 2\epsilon,$$

by definition of n_o. Thus we have shown $\liminf_{n\to\infty} \phi(n) \geq \beta - 2\epsilon$, as desired. Let ϵ tend to zero to complete the proof of (11.8).

11.3. Closeness

The relation (11.8) gives the asymptotics for the worst case boundary functions. It is relatively easy to show that (11.8) implies asymptotics for the worst case standard functionals. To obtain (11.4) from (11.8), it suffices to show

$$|\rho_B^p(n) - \rho^p(n)| = o(n^{(d-p)/d})$$

whenever ρ is the worst case TSP, MST, or minimal matching function. This relation resembles the closeness estimate Lemma 3.7, which says that the TSP, MST, and minimal matching functionals are pointwise close to their respective boundary functionals. Whenever ρ is the worst case TSP, MST, or minimal matching function we will show

(11.10) $$|\rho_B^p(n) - \rho^p(n)| \leq C\left(n^{(d-p-1)/(d-1)} \vee \log n\right).$$

We prove (11.10) individually for the MST, TSP, and minimal matching functionals as follows.

(i) *The Worst Case MST Function.* Since $\mu_B^p(n) \leq \mu^p(n)$, it suffices to show

(11.11) $$\mu^p(n) \leq \mu_B^p(n) + C(n^{(d-p-1)/(d-1)} \vee \log n).$$

Let n be given and let $F \subset [0,1]^d$ denote a set of size n which realizes the worst case MST function $\mu^p(n)$. Let T denote the spanning tree given by the MST boundary functional $M_B^p(F)$ on F. Lemma 3.8 tells us that the sum of the pth powers of the lengths of the edges rooted to the boundary by T is bounded by $C(n^{(d-p-1)/(d-1)} \vee \log n)$.

Let $\mathcal{M} \subset \partial[0,1]^d$ denote the points where the rooted edges in T meet the boundary. Consider the length $T^p(\mathcal{M})$ of the optimal TSP tour T' on $\mathcal{M}$; the edges in T' lie on $\partial[0,1]^d$. Using the edges in the rooted tree T as well as those in the tour T' and applying the triangle inequality $(x + y + z)^p \leq C(x^p + y^p + z^p)$ for x, y, and z positive, we obtain a feasible spanning tree through F having a length of at most

(11.12) $$M_B^p(F) + CT^p(\mathcal{M}) + C(n^{(d-p-1)/(d-1)} \vee \log n),$$

where the last term represents a bound on the pth powers of the lengths of the boundary rooted edges, some of which are needed *two times* in the construction of the feasible tree. Since the edges in T' lie on the $(d-1)$-dimensional boundary, we have $T^p(\mathcal{M}) \leq C(n^{(d-p-1)/(d-1)} \vee \log n)$ by Lemma 3.3. The proof of the estimate (11.11) is completed by noting that $\mu^p(n) = M^p(F)$ is bounded by the length of the suboptimal feasible tree given by (11.12) and then using $M_B^p(F) \leq \mu_B^p(n)$.

(ii) *The Worst Case TSP Function.* We need to show the estimate

(11.13) $$\tau^p(n) \leq \tau_B^p(n) + C(n^{(d-p-1)/(d-1)} \vee \log n).$$

Given n, let $F \subset [0,1]^d$ denote a set of size n which realizes the value of the worst case TSP function $\tau^p(n)$. Consider the boundary functional $T_B^p(F)$. Let T be the tour given by this functional and let $F' \subset F$ denote the subset of F which is rooted to the boundary by the tour T. Let $\mathcal{M} \subset \partial[0,1]^d$ denote the set of points where the edges meet the boundary. Let N denote the common cardinality of the sets F' and $\mathcal{M}$. The goal here is to use T to construct a feasible tour through F.

Consider the length $S^p(\mathcal{M})$ of the minimal matching on $\mathcal{M}$ with edges on $\partial[0,1]^d$. This matching generates tours $C_1, ..., C_R$ ($R \leq N$) on the union $F \cup \mathcal{M}$. Given tour C_i, $1 \leq i \leq R$, select a point $M_i \in \mathcal{M} \cap C_i$ and set $\mathcal{M}' := (M_1, ..., M_R)$. The triangle inequality, the estimate $S^p(\mathcal{M}) \leq C(n^{(d-p-1)/(d-1)} \vee 1)$, and Lemma 3.8 (for the TSP) together tell us that we may add and delete edges from the tours $C_1, ..., C_R$ to generate tours $C_1', ..., C_R'$ on the smaller set $F \cup \mathcal{M}'$ at an extra cost bounded by $C(n^{(d-p-1)/(d-1)} \vee \log n)$. Moreover, the sum of the pth powers of the lengths of the edges with a vertex in $\mathcal{M}'$ is bounded by $C\left(n^{(d-p-1)/(d-1)} \vee \log n\right)$.

Finally, consider the optimal tour of length $T^p(\mathcal{M}')$ with edges which lie on the boundary of $[0,1]^d$. By Lemma 3.3, $T^p(\mathcal{M}') \leq C(n^{(d-p-1)/(d-1)} \vee 1)$.

The above construction, which is achieved at a cost of at most $T_B^p(F) + C(n^{(d-p-1)/(d-1)} \vee \log n)$, generates a connected graph G through $F \cup \mathcal{M}'$ consisting of tours $C_1', ..., C_R'$ through $F \cup \mathcal{M}'$ as well as single tour through $\mathcal{M}'$ with length at most $C(n^{(d-p-1)/(d-1)} \vee 1)$. Since the sum of the pth powers of the lengths of the edges in G with a vertex in $\mathcal{M}'$ is bounded by $C(n^{(d-p-1)/(d-1)} \vee \log n)$, the triangle inequality and an obvious patching argument imply that we may construct a tour through F at an extra cost of at most $C(n^{(d-p-1)/(d-1)} \vee \log n)$. We have thus shown

$$\tau^p(n) \leq T_B^p(F) + C(n^{(d-p-1)/(d-1)} \vee \log n),$$

Since $T_B^p(F) \leq \tau_B(n)$, (11.13) follows as desired.

(iii) *The Worst Case Minimal Matching Function.* We may show

(11.14) $$\sigma^p(n) \leq \sigma_B^p(n) + C(n^{(d-p-1)/(d-1)} \vee \log n)$$

by following the arguments used to prove the analogous estimate (11.11). To prove (11.14), let $F \subset [0,1]^d$ denote a set of size n which realizes the worst case minimal matching $\sigma^p(n)$. Let T denote the graph described by $S_B^p(F)$ and let $\mathcal{M} \subset \partial[0,1]^d$ the set of points where the edges in T meet the boundary. Let $F' \subset F$ be the set of points in F which are matched to the boundary. Consider the length $S^p(\mathcal{M})$ of the minimal matching on $\mathcal{M}$ with edges lying on $\partial[0,1]^d$. By Lemma 3.3, the edges given by $S^p(\mathcal{M})$ have a total length of at most $C(n^{(d-p-1)/(d-1)} \vee 1)$. Using these edges we may construct a natural pairing of points in F' which by the triangle inequality and Lemma 3.8 applied to $S_B^p(F)$, is achieved at an extra cost of at most $C(n^{(d-p-1)/(d-1)} \vee \log n)$. This produces a feasible matching of F and shows that

$$\sigma^p(n) \leq S_B^p(F) + +C(n^{(d-p-1)/(d-1)} \vee \log n).$$

Since $S_B^p(F) \leq \sigma_B^p(n)$, the proof of (11.14) is complete.

We have thus established that the worst case versions of the TSP, MST, and minimal matching functions are close to their worst case boundary versions in the sense that the estimate (11.10) is satisfied. Since the worst case boundary versions satisfy the asymptotics (11.4) the proof of Theorem 11.2 is complete.

11.4. Concluding Remarks

1. It is unclear whether the present method yields rates of convergence with an error term which is more precise than $C(n^{(d-p-1)/(d-1)} \vee \log n)$. Using superadditivity of the boundary functional together with (11.10) it is straightforward to obtain the upper bound

$$\tau^p(n) \leq \beta(\tau^p, d) n^{(d-p)/d} + C(n^{(d-p-1)/(d-1)} \vee \log n)$$

with similar estimates for $\mu^p(n)$ and $\sigma^p(n)$. It is unclear whether this can be developed into a two-sided inequality.

2. It is also unclear whether Theorem 11.2, which holds for power weighted edges with power p satisfying $1 \leq p < d$, can be modified to treat powers p lying in the ranges $0 < p < 1$ and $d \leq p < \infty$. In Yukich (1996a) it was claimed that Theorem 11.2 holds for all $0 < p < 1$, but this remains to be shown.

3. The worst case functions all satisfy the "smoothness" condition

$$|\rho^p(n) - \rho^p(n+k)| \leq C k^{(d-p)/d},$$

where ρ denotes either the TSP, MST, or minimal matching function. It is not clear whether this smoothness property can be put to good use.

4. Steele and Snyder (1989) and Snyder (1987) were the first to investigate the asymptotics of the worst case functions (11.1) and (11.2). Instead of using boundary functionals they obtain asymptotics for $\tau^1(n)$ and $\mu^1(n)$ by showing that these functions have a slow incremental rate of growth and an approximate recursiveness which is akin to a superadditivity condition with no error term. They essentially show that if ρ denotes the worst case version of either the TSP or MST, then $\rho^p(n+1) \leq \rho^p(n) + C n^{-p/d}$. From this recursion we immediately deduce that ρ satisfies the smoothness condition mentioned in Remark 3. Theorem 11.2 extends upon Steele and Snyder (1989) by treating the general case of power-weighted edges (i.e., $p > 1$).

5. There has been considerable work on estimating the constants $\beta(\rho^p, d)$ for various choices of the functional ρ. Steele and Snyder (1989) provides historical background on this subject and forms the basis for the remarks here. In dimension 2, Fejes-Tóth (1940) showed the lower bound $\tau(n) \geq (1-\epsilon)(4/3)^{1/4}n^{1/2}$ for all $n \geq n_o(\epsilon)$. Still in dimension 2, Verblunsky (1951) showed that $\tau(n) \leq (2.8n)^{1/2} + 3.15$ and later this was improved by Few (1955) to $\tau(n) \leq (2n)^{1/2} + 1.75$. For general $d \geq 2$, Few (1955) showed that $\tau(n) \leq d(2(d-1))^{(1-d)/2d} n^{(d-1)/d} + O(n^{1-2/d})$.

More recently, Supowit, Reingold, and Plaisted (1983) showed the lower bound $\tau(n) \geq (4/3)^{1/4}n^{1/2}$ for all $n \geq 1$. Finally, Karloff (1987) showed that in dimension 2 one has $\tau(n) < 0.984(2n)^{1/2} + 11$.

6. Snyder and Steele (1990) treat the asymptotics for the worst case version of the greedy matching heuristic. It remains to be seen whether the methods of this chapter deliver asymptotics for this and other heuristics.

7. Most of this chapter is based on Yukich (1996a).

REFERENCES

Ajtai, M., J. Komlós, and G. Tusnády (1984) On optimal matchings, Combinatorica, 4, 259-264.

Aldous, D. and J. M. Steele (1992) Asymptotics for Euclidean minimal spanning trees on random points, Prob. Theory and Related Fields, 247-258.

Aldous, D. and J. M. Steele (1993) Introduction to the interface of probability and algorithms, Statistical Science, 8, 3-9.

Alexander, K. (1994) Rates of convergence of means for distance-minimizing subadditive Euclidean functionals, Ann. Appl. Prob. 4, 902-992.

Alexander, K. (1996) The RSW theorem for continuum percolation and the CLT for Euclidean minimal spanning trees, Ann. Appl. Prob., 6, 466-494.

Alon, N., J. H. Spencer, and P. Erdös (1992) The Probabilistic Method, Wiley-Interscience Series in Discrete Mathematics and Optimization, Wiley and Sons.

Avis, D. (1983) A survey of heuristics for the weighted matching problem, Networks, 13, 475-493.

Avis, D., B. Davis, and J. M. Steele (1988) Probabilistic analysis of a greedy heuristic for Euclidean matching, Probability in the Engineering and Information Sciences, 2, 143-156.

Avram, F. and D. Bertsimas (1992) The minimum spanning tree constant in geometric probability and under the independent model: A unified approach, Ann. Appl. Prob., 2, 113-130.

Avram, F. and D. Bertsimas (1993) On central limit theorems in geometrical probability, Ann. Appl. Prob., 3, 1033-1046.

Azuma, K. (1967) Weighted sums of certain dependent random variables, Tohuku Math. Journal, 19, 357-367.

Akcoglu, M. A. and U. Krengel (1981) Ergodic theorems for superadditive processes, J. Reine Ang. Math., 323, 53-67.

Baldi, P. and J. Rinott (1989) On normal approximation of distributions in terms of dependency graphs, Ann. Prob., 17, 1646-1650.

Balinski, M. L. (1965) Integer programming: methods, uses, and computation, Management Science, 12, 253-313.

Barahona, F., R. Maynard, R. Rammal and J. P. Uhry (1982) Morphology of ground states of two dimensional frustration model, J. Phys. A: Math. Gen. 15, 673-699.

Beardwood, J., J. H. Halton, and J. M. Hammersley (1959) The shortest path through many points, Proc. Camb. Philos. Soc., 55, 299-327.

Bern, M. and D. Eppstein (1992) Mesh generation and optimal triangulation, in Computing in Euclidean Geometry by F. K. Hwang and D. Z. Du, editors, World Scientific.

Bertsimas, D. and G. van Ryzin (1990) An asymptotic determination of the minimum spanning tree and minimum matching constants in geometrical probability, Oper. Research Lett., 9, 223-231.

Bickel, P. J. and L. Breiman (1983) Sums of functions of nearest neighbor distances, moment bounds, limit theorems, and a goodness of fit test, Ann. Prob., 11, 185-214.

Bieche, I., R. Maynard, R. Rammel and J.P. Uhry (1980) On the ground states of the frustration model of a spin glass by a matching method of graph theory, J. Phys. A: Math. Gen. 13, 2553-2576.

Bollabás, B. (1985) Random Graphs, Academic Press, New York.

Bonomi E. and J. L. Lutton (1984) The N-City traveling salesman problem: statistical mechanics and the Metropolis algorithm, SIAM Review, 26, 551- 568.

Borovkov, A. A. (1962) A probabilistic formulation of two economic problems, Dokl. Acad. Nauk. SSSR, 146, 5, 983-986; English translation in Sov. Math. Dokl., 3, 5, 1403-1406.

Boruvka, O. (1926) O jistém problému minimálnim, Práce Mor. Přírodověd. Spol v Brně (Acta Societ. Scient. Natur. Moravicae) 3, 37-58 (Czech) (= On a Minimal Problem, Prace Morawske Predovedecke Spolecnosti, 3, 37-58).

Christofides, N., A. Mingozzi, P. Toth, and C. Sandi (eds) (1979) Combinatorial Optimization, John Wiley and Sons.

Coffman, E. G. and George Lueker (1991) Probabilistic Analysis of Packing and Partitioning Algorithms, Wiley Interscience Series in Discrete Mathematics and Optimization.

Courant, R. (1950) Dirichlet's Principle, Conformal Mapping, and Minimal Surfaces, Interscience Publications.

Courant, R. and H. Robbins (1941) What is Mathematics?, Oxford University Press.

Dembo, A. and O. Zeitouni (1993) Large Deviations Techniques and Applications, Jones and Bartlett.

Deuschel, Jean-Dominique and Daniel W. Stroock (1989) Large Deviations, Academic Press.

Dobrić, V. and J. E. Yukich (1995) Exact asymptotics for transportation cost in high dimensions, J. Theoretical Probab., 97-118.

Douglas, J. (1939) Minimal surfaces of higher topological structure, Ann. Math., 40, 205-298.

Dunford, N. (1951) An individual ergodic theorem for noncommutative transformations, Acta. Sci. Math. (Szeged), 14, 1-4.

Edmonds, J. (1965) Paths, trees, and flowers, Canadian J. Math. 17, 449-467.

Fejes-Tóth, L. T. (1940) Über einen geometrischen Satz, Math. Z., 46, 83-85.

Few, L. (1955) The shortest path and the shortest road through n points in a region, Mathematika, 2, 141-144.

Few, L. (1962) Average distances between points in a square, Mathematika, 9, 111-114.

Fisher, M. L. and D. S. Hochbaum (1980) Probabilistic analysis of the planar K-median problem, Math. Oper. Research 5, 27-34.

Garey, M. R. and D. S. Johnson (1979) Computers and Intractibility: A Guide to the Theory of NP-Completeness, Freeman, San Francisco.

Gilbert, E. N. (1965) Random minimal trees, SIAM J. Appl. Math., 13, 376-387.

Gilbert, E. N. and H. O. Pollak (1968) Steiner random minimal trees, SIAM J. Appl. Math., 16, 1-29.

Goemans, M. and D. Bertsimas (1991) Probabilistic analysis of the Held-Karp relaxation for the Euclidean traveling salesman problem, Math. Oper. Research, 16, 72-89.

Grimmett, G. R. (1976) On the number of clusters in the percolation model, J. London Math. Soc., 13, 346-350.

Haimovich, M. and A. H. G. Rinnooy Kan (1985) Bounds and heuristics for capacitated routing problems, Math. Oper. Research, 10, 4, 527-542.

Haimovich, M., A. H. G. Rinnooy Kan, and L. Stougie (1988) Analysis of heuristics for vehicle routing problems, in: Vehicle Routing: Methods and Studies, eds. B. Golden and A. Assad, Studies in Management Science and Systems 16, North-Holland.

Halton, J. H. and R. Terada (1982) A fast algorithm for the Euclidean traveling salesman problem, optimal with probability one, SIAM J. Comput. 11, 28-46.

Hammersley, J. M. (1974) Postulates for subadditive processes, Ann. Prob. 2, 652-680.

Held, M. and R. M. Karp (1970) The traveling salesman and minimal spanning trees, Operations Research 18, 1138-1162.

Held, M. and R. M. Karp (1971) The traveling salesman problem and minimal spanning trees: Part II, Math. Progr., 1, 6-25.

Hille, E. (1948) Functional Analysis and Semi-groups, Amer. Math. Soc. Colloquium Publications, no. 31, New York.

Hochbaum, D. and J. M. Steele (1982) Steinhaus's geometric location problem for random samples in the plane, Advances in Appl. Prob., 14, 55-67.

Jaillet, P. (1992) Rates of convergence for quasi-additive smooth Euclidean functionals and applications to combinatorial optimization problems, Math. Oper. Research, 17, 965-980.

Jaillet, P. (1993a) Rate of convergence for the Euclidean minimum spanning tree law, Oper. Research Letters, 14, 73-78.

Jaillet, P. (1993b) Cube versus torus models and Euclidean minimum spanning tree constant, Ann. Appl. Prob., 3, 582-592.

Jaillet, P. (1993c) Analysis of probabilistic combinatorial optimization problems in Euclidean spaces, Math. Oper. Research, 18, 51-70.

Jaillet, P. (1985) On properties of geometric random problems in the plane, Annals of Oper. Research, 61, 1-20.

Jessen, R. J. (1942) Statistical investigation of a sample farm survey, Bull. Iowa St. Coll. Agric. Res. 304.

Karloff, H. J. (1987) How long can a Euclidean traveling salesman tour be?, Tech. Report, Dept. of Computer Science, University of Chicago, Chicago, IL.

Karp, R. M. (1976) The probabilistic analysis of some combinatorial search algorithms, J. F. Traub (ed), Algorithms and Complexity: New Directions and Recent Results, Academic Press, New York, 1-19.

Karp, R. M. (1977) Probabilistic analysis of partitioning algorithms for the traveling salesman problem in the plane, Math. Oper. Research, 2, 209-224.

Karp, R. M. and J. M. Steele (1985) Probabilistic analysis of heuristics, in The Traveling Salesman Problem: A Guided tour of Combinatorial Optimization (Lawler, E.L. et al. eds.) John Wiley and Sons, New York, 181-206.

Kesten, H. and S. Lee (1996) The central limit theorem for weighted minimal spanning trees on random points, Ann. Appl. Prob., 6, 495-527.

Kingman, J. F. C. (1968) The ergodic theory of subadditive stochastic processes, J. Royal Statist. Soc., Ser. B, 30, 499-510.

Kingman, J. F. C. (1973) Subadditive ergodic theory, Ann. Prob., 1, 883-909.

Kirsch, W. and F. Martinelli (1982) On the density of states of Schrodinger operators with a random potential, J. Phys. A.: Math. Gen. 15, 2139-2156.

Krengel, U. (1985) Ergodic Theorems, de Gruyter Studies in Mathematics 6.

Krengel, U. and R. Pyke (1987) Uniform pointwise ergodic theorems for classes of averaging sets and multiparameter processes, Stoch. Processes and Their Applications, 26, 289-296.

Kruskal, J. B. (1956) On the shortest spanning subtree of a graph and the traveling salesman problem, Proc. Amer. Math. Soc., 7, 48-50.

Kuhn, H. W. (1974) Steiner's problem revisited, in Studies in Optimization, G. Dantzig and B.C. Eaves (eds), MAA, Washington, D.C., 52-70.

Lalley, S. P. (1990) Traveling salesman with a self-similar itinerary, Prob. in Engineering and Inform. Sciences, 4, 1-18.

Lawler, E. L., J. K. Lenstra, A. H. G. Rinnooy Kan, D. B. Schmoys (1985) The Traveling Salesman Problem, Wiley Interscience.

Ledoux, M. (1996) Isoperimetry and Gaussian Analysis, Ecole d'été de Probabilités de Saint-Flour (1994), Lecture Notes in Mathematics, 1648, Springer-Verlag.

Ledoux, M. and M. Talagrand (1991) Probability in Banach Spaces, Springer-Verlag.

Lee, S. (1997a) The central limit theorem for Euclidean minimal spanning trees I, Ann. Appl. Prob., to appear.

Lee, S. (1997b) The central limit theorem for Euclidean minimal spanning trees II, preprint.

Lovász, L. and M. D. Plummer (1986) Matching Theory, Akadémiai Kiadó and North Holland.

Lyons, R. and Y. Peres (1997) Probability on Trees, book in preparation.

Mahalanobis, P. C. (1940) A sample survey of the acreage under jute in Bengal, Sankhya, 4, 511-531.

Marks, E. S. (1948) A lower bound for the expected travel among m random points, Ann. Math. Stat., 19, 419-422.

McElroy, K. (1997) Ph.D. Thesis, Department of Mathematics, Lehigh University, Bethlehem, Pa.

McGivney, K. (1997) Ph.D. Thesis, Department of Mathematics, Lehigh University, Bethlehem, Pa.

McGivney, K. and J. E. Yukich (1997a) Asymptotics for Voronoi tessellations, preprint.

McGivney, K. and J. E. Yukich (1997b) Asymptotics for geometric location problems over random samples, preprint.

Melzak, Z. A. (1973) Companion to Discrete Mathematics, John Wiley and Sons.

Mézard, M., G. Parisi and M.A. Virasoro (1987) Spin Glass Theory and Beyond, World Scientific Publ. Co.

Miles, R. (1970) On the homogeneous planar Poisson point process, Math. Biosciences, 85-127.

Milman, V. and G. Schechtman (1986) Asymptotic Theory of Finite Dimensional Normed Spaces, Lecture Notes in Mathematics, 1200, Springer-Verlag.

Nguyen, X. X. (1979) Ergodic theorems for subadditive spatial processes, Z. Wahr. verw. Gebiete, 159-176.

O'hEigeartaigh, M., J. K. Lenstra, and A. H. G. Rinnooy Kan (eds) (1985) Combinatorial Optimization, Annotated Bibliographies, John Wiley and Sons.

Okabe, A., B. Boots, and K. Sugihara (1992) Spatial Tessellations, Concepts and Applications of Voronoi Diagrams, Wiley.

Papadimitriou, C. H. (1978a) The probabilistic analysis of matching heuristics, Proc. of the 15th Allerton Conf. on Communication, Control, and Computing, 368-378.

Papadimitriou, C. H. (1978b) The Euclidean traveling salesman problem is NP-complete, Theoretical Computer Science, 4, 237-244.

Papadimitriou, C. H. (1981) Worst-case and probabilistic analysis of a geometric location problem, SIAM J. Comput., 10, 3, 542-557.

Papadimitriou, C. H. and K. Steiglitz (1982) Combinatorial Optimization: Algorithms and Complexity, Englewood Cliffs, N.J., Prentice Hall.

Papadimitriou, C. H. and U. V. Vazirani (1984) Geometric problems related to the traveling salesman problem, Journal of Algorithms, 5, 231-246.

Penrose, M. D. (1996) The random minimal spanning tree in high dimensions, Annals of Prob., 24, 1903-1925.

Penrose, M. D. (1997) The longest edge of the random minimal spanning tree, Ann. Appl. Prob., 7, 340-361.

Preparata, Franco P. and Michael I. Shamos (1985) Computational Geometry, Springer Verlag.

Prim, R. C. (1957) Shortest connection networks and some generalizations, BSTJ, 36, 1389-1401.

Rédei, L. (1934) Ein Kombinatorischer Satz, Acta Szeged, 7, 39-43.

Redmond, C. (1993) Boundary rooted graphs and Euclidean matching algorithms, Ph.D. thesis, Dept. of Mathematics, Lehigh University, Bethlehem, PA.

Redmond, C. and J. E. Yukich (1994) Limit theorems and rates of convergence for subadditive Euclidean functionals, Ann. Appl. Prob., 1057-1073.

Redmond, C. and J. E. Yukich (1996) Asymptotics for Euclidean functionals with power weighted edges, Stochastic Processes and Their Applications, 61, 289-304.

Reingold, E. M. and R. E. Tarjan (1981) On a greedy heuristic for complete matching, SIAM Journal of Computing, 10, 676-681.

Rhee, W. (1991) On the fluctuations of the stochastic traveling salesperson problem, Math. Oper. Research, 16, 3, 482-489.

Rhee, W. (1992) On the traveling salesperson problem in many dimensions, Random Structures and Algorithms, 3, 227-233.

Rhee, W. (1993a) On the stochastic Euclidean traveling salesperson problem for distributions with unbounded support, Math. Oper. Research, 18, 292-299.

Rhee, W. (1993b) A matching problem and subadditive Euclidean functionals, Ann. Appl. Prob., 3, 794-801.

Rhee, W. (1994a) Boundary effects in the traveling salesperson problem, Oper. Research Letters, 16, 19-25.

Rhee, W. (1994b) On the fluctuations of simple matching, Oper. Research Letters, 16, 27-32.

Rhee, W. (1994c) Probabilistic analysis of a capacitated vehicle routing problem II, Ann. Appl. Prob., 4, 741-764.

Rhee, W. and M. Talagrand (1987) Martingale inequalities and NP-complete problems, Math. Oper. Research, 12, 177-181.

Rhee, W. and M. Talagrand (1989a) Martingale inequalities, interpolation, and NP-complete problems, Math. Oper. Research, 14, 91-96.

Rhee, W. and M. Talagrand (1989b) A sharp deviation inequality for the stochastic traveling salesman problem, Ann. Prob., 17, 1-8.

Rhee, W. and M. Talagrand (1992) On the long edges in the shortest tour through n random points, Combinatorica, 12, 323-330.

Rosenkrantz, D. J., E. Stearns, and P. M. Lewis (1977) An analysis of several heuristics for the traveling salesman problem, SIAM J. Comput. 6, 563-581.

Salzberg, S., A. Delcher, D. Heath, and S. Kasif (1991) Learning with a helpful teacher, Proc. 12th Int. Joint Conf. Artificial Intelligence.

Santalo, L. A. (1976) Integral Geometry and Geometric Probability, Addison-Wesley.

Shor, P. and J. E. Yukich (1991) Minimax grid matching and empirical measures, Ann. Prob., 1338-1348.

Shorack, G. R. and J. A. Wellner (1986) Empirical Processes with Applications to Statistics, Wiley, New York.

Smith, W. D. (1988) Studies in Computational Geometry Motivated By Mesh Generation, Ph.D. Thesis, Princeton University.

Smythe, R. T. (1976) Multiparameter sub-additive processes, Ann. Prob., 4, 772-782.

Snyder, T. L. (1987) Asymptotic worst case lengths in some problems from classical computational geometry and combinatorial optimization, Ph.D. thesis, Princeton University, Princeton N.J.

Snyder, T. L. (1992) Worst-case minimum rectilinear Steiner trees in all dimensions, Discrete Comput. Geom. 8, 73-92.

Snyder, T. L. and J. M. Steele (1990) Worst-case greedy matchings in the unit d-cube, Networks, 20, 779-800.

Steele, J. M. (1981a) Subadditive Euclidean functionals and non-linear growth in geometric probability, Ann. Prob., 9, 365-376.

Steele, J. M. (1981b) Complete convergence of short paths in Karp's algorithm for the TSP, Math. Oper. Research, 374-378.

Steele, J. M. (1982) Optimal triangulation of random samples in the plane, Ann. Prob., 10, 548-553.

Steele, J. M. (1986) Probabilistic algorithm for the directed traveling salesman problem, Math. Oper. Research, 11, 343-350.

Steele, J. M. (1988) Growth rates of Euclidean minimal spanning trees with power weighted edges, Ann. Prob., 16, 1767-1787.

Steele, J. M. (1990a) Seedlings in the theory of shortest paths, in Disorder in Physical Systems: A volume in Honor of J. M. Hammersley (G. Grimmett and D. Welsh, eds.) Cambridge University Press, 277-306, London.

Steele, J. M. (1990b) Probabilistic and worst case analyses of classical problems of combinatorial optimization in Euclidean space, Math. Oper. Research, 15, 749-770.

Steele, J. M. (1992) Euclidean semi-matching of random samples, Mathematical Programming, 53, 127-146.

Steele, J. M. (1993) Probability and problems in Euclidean combinatorial optimization, Statistical Science, 8, 48-56.

Steele, J. M. (1997) Probability Theory and Combinatorial Optimization, SIAM.

Steele, J. M. and T. L. Snyder (1989) Worst case growth rates of some classical problems of combinatorial optimization, Siam J. Computation, 18, 278-287.

Steinhaus, H. (1956) Sur la division de corps matériels en parties, Bull. Acad. Polon. Sci., 4, 801-804.

Stoyan, D., W. Kendall, and J. Mecke (1995) Stochastic Geometry and Its Applications, Second Edition, Wiley Series in Probability and Statistics.

Supowit, K. J., E. M. Reingold, and D. A. Plaisted (1983) The traveling salesman problem and minimum matchings in the unit square, SIAM J. Comput., 12, 144-156.

Talagrand, M. (1989) Isoperimetry and integrability of the sum of independent Banach space valued random variables, Ann. Prob., 17, 1546-1570.

Talagrand, M. (1991) Complete convergence of the directed TSP, Math. Oper. Research, 881-887.

Talagrand, M. (1992) Matching random samples in many dimensions, Ann. Prob., 2, 846-856.

Talagrand, M. (1994a) Matching theorems and empirical discrepancy computations using majorizing measures, J. Amer. Math. Soc., 7, 455-537.

Talagrand, M. (1994b) The transportation cost from the uniform measure to the empirical measure in dimension ≥ 3, Ann. Prob., 22, 919-959.

Talagrand, M. (1994c) Sharper bounds for Gaussian and empirical processes, Ann. Prob., 22, 28-76.

Talagrand, M. (1995) Concentration of measure and isoperimetric inequalities in product spaces, Publications Mathématiques de l'I. H. E. S., 81, 73-205.

Talagrand, M. (1996a) A new look at independence, Ann. Prob., 24, 1-34.

Talagrand, M. (1996b) New concentration inequalities in product spaces, Invent. Math., 126, 505-563.

Talagrand, M. (1996c) Private communication.

Talagrand, M. and J. E. Yukich (1993) The integrability of the square exponential transportation cost, Ann. Appl. Prob., 3, 1100-1111.

Tarjan, R. E. (1983) Data Structures and Network Algorithms, SIAM.

Van Enter, A. C. D. and J. L. Van Hemmen (1983) The thermodynamic limit for long range systems, J. Statist. Physics 32, 141-152.

Verblunsky, S. (1951) On the shortest path through a number of points, Proc. Amer. Math. Soc., 2, 904-913.

Voronoi, G. (1908) Nouvelles applications des paramètres continus à la théorie des formes quadratiques, J. Reine Angew. Math. 134, 198-287.

Weide, B. (1978) Statistical Methods in Algorithm Design and Analysis, Ph.D. Thesis, Computer Science Department, Carnegie Mellon University.

Wiener, N. (1939) The ergodic theorem, Duke Math. Journal, 5, 1-18.

Yukich, J. E. (1989) Optimal matching and empirical measures, Proc. Amer. Math. Soc., 1051-1059.

Yukich, J. E. (1992) Some generalizations of the Euclidean two-sample matching problem, Progress in Probability, Birkhauser, vol. 30, 55-66.

Yukich, J. E. (1995a) Asymptotics for the stochastic TSP with power weighted edges, Prob. Theory and Related Fields, 203-220.

Yukich, J. E. (1995b) Quasi-additive Euclidean functionals, Proc. of the IMA Workshop on Probability and Algorithms, 149-158.

Yukich, J. E. (1996a) Worst case growth rates for some classical optimization problems, Combinatorica, 16, 4, 575-586.

Yukich, J. E. (1996b) Ergodic theorems for some classical optimization problems, Ann. Appl. Prob., 6, 3, 1006-1023.

Yukich, J. E. (1997a) Minimal triangulations resemble minimal tours, preprint.

Yukich, J. E. (1997b) Asymptotics for weighted minimal spanning trees on random points, preprint.

Zygmund, A. (1951) An individual ergodic theorem for non-commutative transformations, Acta. Sci. Math. (Szeged), 14, 103-110.

INDEX